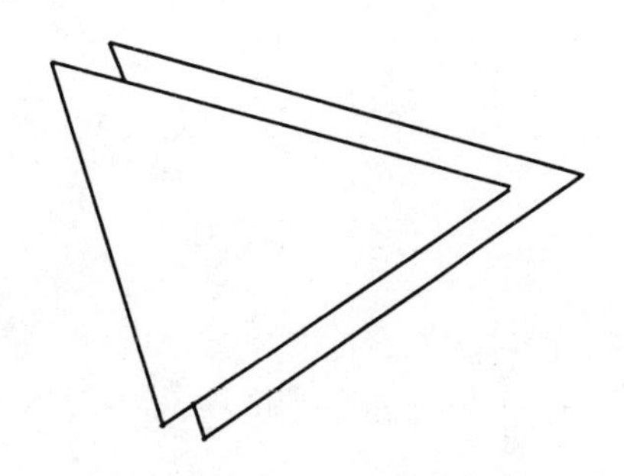

Milady's Standard Practical Workbook

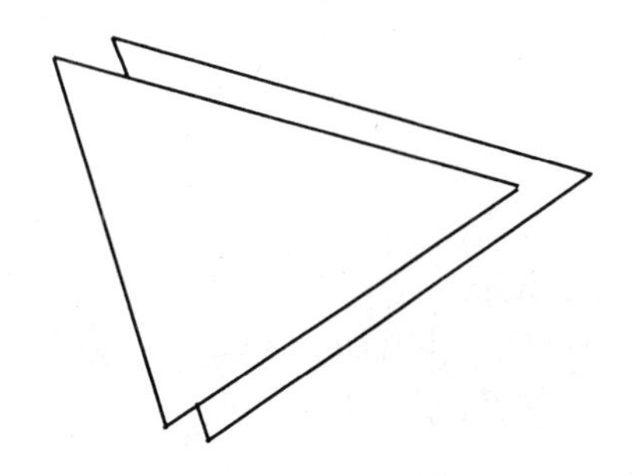

Milady's
Standard Practical Workbook

To be used with
MILADY'S STANDARD TEXTBOOK OF COSMETOLOGY

Compiled by Kathleen Ann Bergant,
Milwaukee Area Technical College

Milady Publishing Company
(A Division of Delmar Publishers, Inc.)
2 Computer Drive West, Box 12519
Albany, New York 12212

Cover credit: Michael A. Gallitelli

Contributors to Cover Art: Linda Balhorn
Steven Landis
Garland Drake International
Tammy Bigan

Photographer: Michael A. Gallitelli, on location at the Austin Beauty School, with Dino Petrocelli

Artists: Edward Tadiello
Robert Richards
Judy Francis
Shizuko Horii
Ron Young

ISBN: 1-56253-007-0

Library of Congress Catalog Card Number: 90-13361

Printed in the United States of America

10 9 8 7 6 5

Contents

How To Use This Workbook

Milady's Standard Practical Workbook has been written to meet the needs, interests, and abilities of students receiving training in cosmetology.

This workbook should be used together with *Milady's Standard Textbook of Cosmetology* and *Milady's Standard Theory Workbook*. This book directly follows the practical information found in the student textbook. Pages to be read and studied are listed at the beginning of each chapter. The theoretical information can be found in *Milady's Standard Theory Workbook*.

Students are to answer each item in this workbook with a pencil after consulting their textbook for correct information. Items can be corrected and/or rated during class or individual discussions, or on an independent study basis.

Various tests are included to emphasize essential facts found in the textbook and to measure the student's progress. "Word Reviews" are listed for each chapter. They are to be used as study guides, for class discussions, or for the teacher to assign groups of words to be used by the student in creative essays.

YOUR PROFESSIONAL IMAGE

See *Milady's Standard Theory Workbook.*

BACTERIOLOGY

See *Milady's Standard Theory Workbook.*

STERILIZATION AND SANITATION

See *Milady's Standard Theory Workbook.*

Date ______________

Rating ______________

Text Pages 43–66

PROPERTIES OF THE SCALP AND HAIR

SCALP CARE

1. Explain how scalp and hair are kept clean.

____

2. Explain why scalp and hair must be kept clean.

____

____

SCALP MANIPULATIONS

3. Explain why scalp manipulations are given with a continuous, even motion.

____

4. Specify how often scalp treatments are given.

____

____

____

5. Explain why cosmetologists should know the muscles and locations of blood vessels and nerve points.

____

____

6. Describe why hands are placed under the hair with each massage movement.

____

____

____

7. Label the following illustrations by filling in the correct names of the movements.

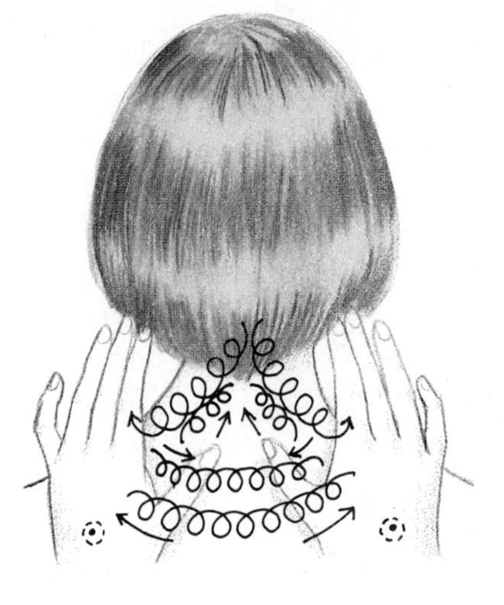

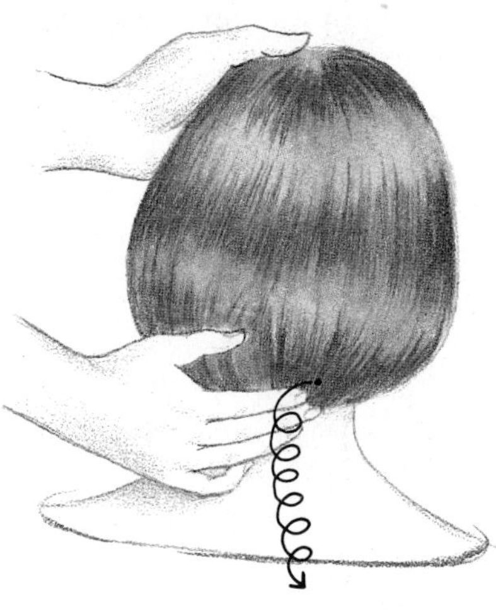

DISORDERS OF THE SCALP

8. Describe a natural occurrence that is commonly mistaken for dandruff.

____

____

DANDRUFF

9. Describe dandruff.

____

10. Identify the medical term for dandruff. ____________

11. Name what can result from long neglected, excessive dandruff.

____

12. Name a direct cause of dandruff.

____

____

13. List six indirect or associated causes of dandruff.

    1. ____________________
    2. ____________________
    3. ____________________
    4. ____________________
    5. ____________________
    6. ____________________

14. List two contributing causes of dandruff.

 1. ____________________

 2. ____________________

15. Give the medical names for the dry type of dandruff and for the greasy or waxy type.

____

____

16. Describe pityriasis capitis simplex.

____

____

____

17. List five ways to treat pityriasis capitis simplex.

 1. ____________________ 2. ____________________

 3. ____________________ 4. ____________________

 5. ____________________

18. Describe pityriasis steatoides.

____

____

____

19. Describe the treatment for pityriasis steatoides. ____________________

20. Explain how both forms of dandruff can be spread.

____

21. Describe how cosmetologists can prevent the spread of dandruff.

____

22. List the eleven steps in the treatment for dandruff.

 1. ____________________

 2. ____________________

 3. ____________________

 4. ____________________

 5. ____________________

 6. ____________________

 7. ____________________

 8. ____________________

 9. ____________________

 10. ____________________

 11. ____________________

ALOPECIA

23. Define alopecia. ______________________________

24. Name the natural occurrence that should not be confused with alopecia.

____

25. Identify the most frequent times of the year for natural shedding of hair.

____

26. Explain what can happen from wearing ponytails and tight braids.

____

27. Describe alopecia senilis.

____

28. Describe alopecia prematura.

____

29. Describe alopecia areata.

____

____

30. Describe the patches associated with alopecia areata.

____

31. Describe the affected areas in alopecia areata.

____

32. Explain the cause of most cases of alopecia areata.

____

33. List the twelve steps in the treatment for alopecia.

    1. ______________________________
    2. ______________________________
    3. ______________________________
    4. ______________________________
    5. ______________________________
    6. ______________________________
    7. ______________________________
    8. ______________________________
    9. ______________________________
    10. ______________________________
    11. ______________________________
    12. ______________________________

34. List the nine steps in the treatment for alopecia areata.

1. ______________________________
2. ______________________________
3. ______________________________
4. ______________________________
5. ______________________________
6. ______________________________
7. ______________________________
____
8. ______________________________
9. ______________________________

VEGETABLE PARASITIC INFECTIONS

35. Define tinea.

____

36. Identify the cause of ringworm. ______________________________

37. List four sources of transmission of ringworm.

1. ______________ 2. ______________
3. ______________ 4. ______________

38. Explain the treatment for any ringworm condition.

____

39. Give the common name for tinea capitis.

____

40. Describe tinea capitis.

____
____
____

41. Give two other names for tinea favosa.

____

42. Describe tinea favosa.

____
____
____

43. Explain why tinea favosa must be treated by a physician.

____

ANIMAL PARASITIC INFECTIONS

44. Describe scabies.

____

45. Identify the cause of scabies. ____________________________

46. List two reasons for the formation of vesicles and pustules.

1. ____________________ 2. ____________________

47. Identify the cause of pediculosis capitis. ____________________

48. Describe how infection can occur.

____

____

49. Explain how pediculosis capitis is transmitted.

____

____

50. Describe how to kill head lice.

____

____

____

____

51. Name two places where head lice should never be treated.

____

STAPHYLOCOCCI INFECTIONS

52. Give the common name for furuncle. ____________________

53. Describe furuncle.

____

____

54. Describe carbuncle.

____

55. Explain how to treat a client with a carbuncle.

____

GENERAL HAIR TREATMENTS

56. Explain the purpose of a general scalp treatment.

____

57. Name a benefit of regular scalp treatments. ____________________

58. List the ten steps in the treatment for normal hair and scalp.

1. ________________
2. ________________
3. ________________
4. ________________
5. ________________
6. ________________
7. ________________
8. ________________
9. ________________
10. ________________

59. Explain when to give a treatment for dry hair and scalp.

____

60. Describe the type of products used to treat dry scalp and hair.

____

61. List the types of products to avoid when treating dry scalp and hair.

1. ________________
2. ________________
3. ________________

62. List the ten steps in the treatment for dry hair and scalp.

1. ________________
2. ________________
3. ________________
4. ________________
   ____
5. ________________
6. ________________
7. ________________
8. ________________
   ____
9. ________________
10. ________________

63. Name the cause of excessive oiliness.

____

64. Explain how to increase blood circulation to the scalp.

____

65. Describe how the correct degree of pressing and squeezing benefits the scalp.

____

66. Explain how to normalize the function of the sebaceous glands.

____

67. List the eleven steps in the treatment for oily hair and scalp.

1. _______________
2. _______________
3. _______________
4. _______________
5. _______________
6. _______________
7. _______________
8. _______________
9. _______________
10. _______________
11. _______________

68. Explain when not to use high-frequency current.

____

69. Identify the purpose of a corrective hair treatment.

____

70. Explain how often hair treatments can be given in relation to various chemical treatments.

____

____

____

71. Describe the type of product used to quickly soften dry hair.

____

72. Explain the reason for applying heat to hair being treated with some conditioners.

___

___

___

73. List the seven steps in the corrective hair treatment.

1. ___
2. ___
3. ___
4. ___
5. ___
6. ___
7. ___

WORD REVIEW

alopecia
corrective
epidermis
fungi
high-frequency
infrared
mite
pediculosis
scabies
sebum
vesicle
carbuncle
dandruff
epithelial
furuncle
horny scales
louse
papules
pityriasis
scutula
staphylococci
contagious
dermatologist
faradic
glass rake electrode
imbrications
manipulations
parasite
pustule
sebaceous
tinea

MATCHING TEST

Insert the correct term or phrase in front of each definition.

alopecia
carbuncle
furuncle
pityriasis capitis simplex
scutula
vesicle
alopecia areata
epidermis
pediculosis capitis
pityriasis steatoides
tinea
alopecia prematura
epithelial
pityriasis
scabies
tinea capitis

1. ______________________________ the greasy or waxy type of dandruff
2. ______________________________ cells located on the skin's surface
3. ______________________________ the technical term for any abnormal hair loss
4. ______________________________ ringworm of the scalp
5. ______________________________ the medical term for dandruff
6. ______________________________ a contagious skin disease caused by the itch mite
7. ______________________________ the medical term for a boil
8. ______________________________ the sudden falling out of hair in round patches
9. ______________________________ the medical term for ringworm
10. ______________________________ the dry type of dandruff
11. ______________________________ the outermost layer of the skin
12. ______________________________ contagious condition caused by the head louse

RAPID REVIEW TEST

Place the correct word in the space provided in each sentence below.

baldness	braids	contact
contagious	cortex	disorders
dry	excessive	follicle
fungi	hair	manipulations
medical	medulla	noncontagious
oil	over	person
ponytails	regular	sebaceous
under	unsanitized	

1. Head lice are transmitted from one ____________ to another by ____________ with infested personal articles.

2. Long neglected, excessive dandruff can lead to ____________.

3. Scalp ________________ are given with all scalp treatments.

4. Excessive oiliness is caused by __________ activity of the ______________ glands.

5. A furuncle is an acute staphylococci infection of a hair ____________.

6. Both forms of dandruff are considered to be ______________.

7. The wearing of ______________ and ____________ can contribute to baldness.

8. Tinea is commonly carried by scales or hairs containing __________.

9. Scalp manipulations are given with hands ___________ the hair.

10. _____________ treatment is advisable for pityriasis steatoides.

11. A healthy, clean scalp will resist a variety of ______________.

12. A deficiency of natural _________ results in _________ scalp and hair.

13. _____________ scalp treatments are beneficial in preventing baldness.

14. ________________ articles are sources of transmission for contagious diseases.

15. Heat helps some conditioners penetrate into the ___________.

MULTIPLE CHOICE TEST

Read each statement carefully, then write the letter representing the word or phrase that correctly completes the statement on the blank line to the right.

1. The natural shedding of the uppermost layer of the scalp should not be mistaken for
   a) alopecia   b) pityriasis
   c) scabies   d) tinea   ________
2. For normal scalps, scalp massage is recommended
   a) once a week   b) twice a month
   c) once a month   d) twice a year   ________
3. With tinea capitis, the patches spread and cause the hair to
   a) become shinier   b) become stronger
   c) break off   d) discolor   ________
4. Contributing causes of dandruff are the use of strong shampoos and insufficient
   a) blotting   b) cutting
   c) draping   d) rinsing   ________
5. The treatment for oily hair and scalp includes manipulating the scalp to increase
   a) flow of sebum   b) flow of perspiration
   c) blood circulation   d) muscular activity   ________
6. Pediculosis capitis is caused by
   a) a round worm   b) a round insect
   c) an animal parasite   d) a vegetable parasite   ________
7. With tinea favosa, the dry, sulfur-yellow, cuplike crusts on the scalp have
   a) a peculiar odor   b) a fragrant aroma
   c) a sweet smell   d) no odor   ________
8. The form of baldness that occurs in old age is
   a) alopecia prematura   b) alopecia senilis
   c) pityriasis steatoides   d) tinea favosa   ________
9. Corrective hair treatments are especially beneficial when given a week before and a week after
   a) a shampoo   b) a blow-dry
   c) a vacation   d) a chemical service   ________
10. All forms of ringworm should be referred to
    a) another cosmetologist   b) another salon
    c) a physician   d) a lawyer   ________
11. A basic requisite for a healthy scalp is
    a) a poor diet   b) poor circulation
    c) strong chemicals   d) cleanliness   ________
12. All scalp treatments include brushing the client's hair for about
    a) 2 minutes   b) 5 minutes
    c) 10 minutes   d) 15 minutes   ________

13. The natural shedding of hair occurs most frequently in

a) spring and summer
b) spring and fall
c) fall and winter
d) summer and winter ________

14. Ringworm is caused by

a) a round worm
b) a round insect
c) an animal parasite
d) a vegetable parasite ________

15. Clients with dry hair and scalp should avoid the use of

a) strong soaps
b) mild soaps
c) moisturizers
d) emollients ________

16. High-frequency current should not be used on hair treated with tonics or lotions having

a) a sweet smell
b) a pungent odor
c) a watery consistency
d) an alcohol content ________

17. A corrective hair treatment deals with the

a) scalp
b) hair shaft
c) hair follicles
d) sebaceous glands ________

18. Pityriasis steatoides is a condition in which the scales become mixed with

a) dust
b) dirt
c) sebum
d) perspiration ________

19. Scalp manipulations that soothe the client's tension are given with a

a) continuous, even motion
b) slow, slapping motion
c) fast, kneading motion
d) fast, slapping motion ________

20. Head lice feed on

a) dead skin cells
b) natural oil
c) the hair
d) the scalp ________

21. The medical term for dandruff is

a) alopecia
b) pityriasis
c) scabies
d) tinea ________

22. Many scalp treatments include placing the client under the infrared lamp for about

a) 2 minutes
b) 5 minutes
c) 10 minutes
d) 15 minutes ________

23. The affected areas in alopecia areata are due to a decreased supply of

a) sebum
b) perspiration
c) blood
d) water ________

24. Products used to soften dry hair usually contain

a) alcohol
b) ammonia
c) chlorine
d) cholesterol ________

25. Scabies is caused by

a) a round worm
b) a round insect
c) an animal parasite
d) a vegetable parasite ________

Also see *Milady's Standard Theory Workbook.*

Date ______________

Rating ______________

Text Pages 67–72

DRAPING

INTRODUCTION

1. Explain why protection of the clients' skin and clothing is necessary.

____

____

____

2. List the six steps necessary before draping a client for any type of service.
   1. ______________________________
   2. ______________________________
   3. ______________________________
   4. ______________________________
   5. ______________________________
   6. ______________________________

3. Explain the purpose of the towel or neck strip.

____

____

DRAPING FOR WET HAIR SERVICES

4. Describe how to place the first towel when draping for wet hair services.

____

____

5. Describe how to position the cape.

____

____

6. Describe how to position the second towel.

____

7. Explain why, in preparation for haircutting, the towel is replaced by a neck strip.

____

DRAPING FOR CHEMICAL SERVICES

8. Describe how to position the towel when draping for chemical services.

____

____

____

9. Describe how to position the cape.

____

10. Describe how to adjust and fold the towel to complete the draping.

____

____

11. Explain the purpose of applying a protective cream around the hairline immediately before applying chemicals to the hair.

____

DRAPING FOR DRY HAIR SERVICES

12. List the three steps required to drape a client for thermal design.
    1. ____

       ____
    2. ____

       ____
    3. ____

       ____

13. Identify what should not touch the client's neck.

____

14. List the two steps required to drape a client for a comb-out.
    1. ____

       ____
    2. ____

       ____

WORD REVIEW

cape
neck strip
skin irritation
draping
obstruction
lengthwise
protective cream

RAPID REVIEW TEST

Place the correct word in the space provided in each sentence below.

after	before	cape
dry	hair	hairline
neck	neck strip	over
sanitary	towel	uncovered
wet		

1. The purpose of the towel or neck strip is for _______________reasons.
2. When draping for chemical services, the cape is placed over the _____________.
3. Objects should be removed from the client's hair _____________draping.
4. When draping for a comb-out, the cape is placed over the ________________.
5. Protective cream is applied around the _______________.
6. When draping for wet hair services, the cape is placed ____________the first towel.
7. For haircutting, the ___________must be allowed to fall naturally without obstruction.
8. When draping for dry hair services, the ________________portion of the neck strip is folded down over the cape.
9. The ____________must not touch the client's skin.
10. When draping for ___________hair services, another towel is placed over the cape.

MULTIPLE CHOICE TEST

Read each statement carefully, then write the letter representing the word or phrase that correctly completes the statement on the blank line to the right.

1. The purpose of draping is to protect the client's
   a) hair b) face
   c) scalp d) skin and clothing ________

2. Before draping, the client's collar is protected by
   a) turning to the inside b) covering with a cloth
   c) turning to the outside d) covering with a plastic bag ________

3. When draping for wet services, the first towel is placed lengthwise across the
   a) client's head b) client's face
   c) client's shoulders d) client's back ________

4. Application of protective cream helps to prevent skin irritation from
   a) water b) conditioners
   c) shampoos d) hair chemicals ________

5. Before draping a client for any type of service, the cosmetologist must sanitize
   a) his/her hands b) the client's scalp
   c) the client's neck d) the client's jewelry ________

6. When draping for dry services, the cape is placed
   a) on the client's skin b) over a neck strip
   c) over a towel d) over a neck strip and towel ________

7. For haircutting, the towel should be removed after shampooing and replaced with
   a) another towel b) a neck strip
   c) a plastic cape d) a cloth cape ________

8. Before draping, the client should remove all jewelry and
   a) store it away b) put it in her lap
   c) put it on the work station d) put it at the shampoo bowl ________

9. When draping for wet services, the second towel is placed
   a) under the cape b) over the first towel
   c) over the cape d) over the neck strip ________

10. When draping for any type of service, it's necessary that the cape
    a) look attractive b) touch the client's skin
    c) look fashionable d) not touch the client's skin ________

Date ____________

Rating ____________

Text Pages 73–82

SHAMPOOING, RINSING, AND CONDITIONING

INTRODUCTION

1. Explain why shampooing is an important preliminary step for other hair services.

____

____

____

2. Name the primary purpose of a shampoo. ____________________

3. Explain what a shampoo must accomplish to be effective.

____

____

4. Explain the purpose of analyzing the client's hair and scalp before the shampoo.

____

5. Describe how to treat a client with an infectious disease.

____

6. Explain how scalp disorders can result from failure to shampoo regularly.

____

____

7. Specify how often hair should be shampooed.

____

8. Name the type of hair that requires more frequent shampooing. ____________

WATER

9. Describe the chemical composition of water. ____________________

10. Describe soft water.

____

____

11. Describe hard water. ____

____

SELECTING THE CORRECT SHAMPOO

12. Explain why cosmetologists should learn the composition and action of a shampoo.

____

13. Explain how cosmetologists can make informed decisions about a shampoo's composition and action.

____

14. List seven hair conditions that are not considered normal.
    1. ____
    2. ____
    3. ____
    4. ____
    5. ____
    6. ____
    7. ____

REQUIRED MATERIALS AND IMPLEMENTS

15. Describe what happens when the cosmetologist fails to gather all necessary items before beginning the shampoo.

____

16. List six materials and implements required for a shampoo.
    1. ____________________ 2. ____________________
    3. ____________________ 4. ____________________
    5. ____________________ 6. ____________________

BRUSHING

17. Describe natural bristles and their recommended use.

____

____

____

18. Describe nylon bristles and their recommended use.

____

19. List two instances when a thorough brushing should not be given.
    1. ____________________ 2. ____________________

20. List three beneficial effects of brushing.

    1. ______________________________
    2. ______________________________
    3. ______________________________

21. Identify the size of hair sections used for brushing. ______________________________

SHAMPOO PROCEDURE

22. List the 12 steps required in the preparation for a shampoo.

    1. ______________________________
    2. ______________________________
    3. ______________________________
    4. ______________________________
    5. ______________________________
    6. ______________________________
    7. ______________________________
    8. ______________________________
    9. ______________________________
    10. ______________________________
    11. ______________________________
    12. ______________________________

23. Describe how to adjust the water temperature.

____

____

24. Explain how to test the water temperature.

____

25. Explain how to monitor the water temperature.

____

26. Name the first step in the shampoo procedure.

____

27. Describe the actions of the free hand while wetting the hair.

____

____

28. Name the second step in the shampoo procedure.

____

29. Identify the area of the head where the application of shampoo is started.

____

30. Name the part of the fingers used to work shampoo into a lather.

____

31. List three instances when firm pressure should not be applied in massaging the scalp.
    1. ____
    2. ____
    3. ____

32. Name the third step in the shampoo procedure. ______________________

33. Describe the type of movement used to manipulate the scalp.

____

34. Identify how to control the client's head while shampooing the back of the head.

____

35. Explain how to give rotary movements after allowing the client's head to relax.

____

36. Describe how to remove excess shampoo and lather. ______________________

37. Name the fourth step in the shampoo procedure.

____

38. Describe how to force the spray of water against the base scalp area.

____

39. Name the fifth step in the shampoo procedure.

____

40. Explain why less shampoo is needed for a second application.

____

41. Name the sixth step in the shampoo procedure. ______

42. Identify what is removed from the hair at the shampoo bowl. ______

43. Describe how to remove excess moisture from the client's face and ears.

______

44. Explain how to finish towel-drying the hair.

______

______

______

45. Explain how to comb hair after the shampoo.

______

46. List the five steps required in the cleanup.

1. ______
2. ______
3. ______

   ______
4. ______
5. ______

SHAMPOOING CHEMICALLY TREATED HAIR

47. Specify the type of shampoo recommended for chemically treated hair.

______

48. Explain how to remove tangles from chemically treated hair.

______

______

WORD REVIEW

bristles
infectious
massage
pressure
rinse
sensitive
temperature
composition
lather
minerals
professionalism
sanitize
shampoo cape
disinfectant
manipulate
moisture
relaxing
saturate
sheen

RAPID REVIEW TEST

Place the correct word in the space provided in each sentence below.

alcohol
chemical
disorders
exposure
inner
normal
outer
stimulate
back
cushions
disinfectant
front
lightened
nozzle
sensitive
temperature
brushing
disease
dry
hydrogen
nitrogen
oily
shampoos

1. Hair is not considered ____________ if it has been damaged by exposure to the elements.
2. Analyzing clients' hair and scalp is necessary to check for disease or ____________.
3. Water is composed of ____________ and oxygen.
4. At the shampoo sink, the shampoo cape is adjusted over the ____________ of the shampoo chair.
5. Hair brushing helps to ____________ blood circulation to the scalp.
6. Shampooing is done with the ____________ of the fingers.
7. Brushing should be eliminated on clients who are about to receive ____________ services.
8. Water temperature is tested by spraying it on the ____________ side of the wrist.
9. Manipulations begin at the ____________ hairline.
10. Combs and brushes are cleaned by removing hair; washing in hot, soapy water; then placing in a wet ____________.
11. Firm pressure should not be used in massaging the scalp if the client's scalp is ____________.
12. Hair is not considered normal if it has been ____________, toned, or tinted.
13. Generally, ____________ hair requires more frequent shampooing than normal or dry hair.
14. Water temperature is monitored constantly by keeping one finger over the edge of the spray ____________.
15. Scalp disorders may result when oil and perspiration mix with natural scales and dirt, offering a breeding place for ____________-producing bacteria.

MULTIPLE CHOICE TEST

Read each statement carefully, then write the letter representing the word or phrase that correctly completes the statement on the blank line to the right.

1. The primary purpose of a shampoo is to
   a) make styling easier   b) treat scalp disorders
   c) make combing easier   d) cleanse the hair and scalp ________

2. Hairbrushes made of natural bristles are recommended for hair
   a) brushing   b) designing
   c) styling   d) braiding ________

3. A shampoo's ability to lather is decreased when used with
   a) soft water   b) hard water
   c) permanent waves   d) hair lighteners ________

4. Shampoo is first applied to the head at the
   a) hairline   b) crown
   c) nape   d) top ________

5. A client with an infectious disease should be referred to
   a) another cosmetologist   b) another salon
   c) a physician   d) his/her relatives ________

6. Before rinsing, excess shampoo and lather are removed from hair by
   a) blow-drying   b) wiping
   c) blotting   d) squeezing ________

7. After shampooing, used towels are placed
   a) on the floor   b) on the work station
   c) in a corner   d) in the towel hamper ________

8. The partings used to brush hair should measure about
   a) 1/4"   b) 1/2"
   c) 1"   d) 2" ________

9. When adjusting water temperature, cosmetologists should first turn on the
   a) cold water   b) tepid water
   c) warm water   d) hot water ________

10. Chemically treated hair requires
    a) an alkaline shampoo   b) a medicated shampoo
    c) a mild shampoo   d) a harsh shampoo ________

Date ______________

Rating ______________

Text Pages 83–100

HAIRCUTTING

INTRODUCTION

1. Explain the importance of a good haircut.

____

____

2. Describe the purpose of hairstyles.

____

____

3. List seven factors that must be considered when selecting a suitable hairstyle.

1. ______________ 2. ______________
3. ______________ 4. ______________
5. ______________ 6. ______________
7. ______________

IMPLEMENTS USED IN HAIRCUTTING

4. Label the parts of the haircutting scissors.

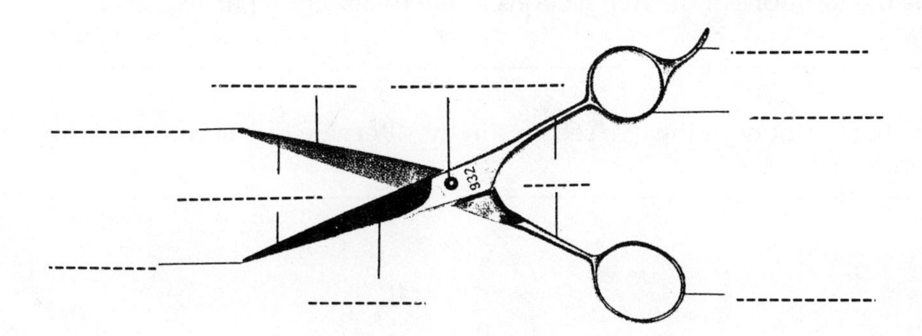

5. List six implements used in haircutting.

    1. ______________________________

    2. ______________________________

    3. ______________________________

    4. ______________________________

    5. ______________________________

    6. ______________________________

SECTIONING FOR HAIRCUTTING

6. Label the parts of the straight razor.

7. Name the locations of the four sections in the four-section parting.

____

8. Name the locations of the five sections in the five-section parting.

____

9. Name the locations of the five sections in the alternate five-section parting.

____

10. Number the sections in the order in which they are parted off for haircutting.

HOLDING HAIR SHAPING IMPLEMENTS

11. Describe how to correctly handle the haircutting scissors.

____

____

____

12. Describe the purpose of thinning shears. ______________________

13. Name the type of thinning shear that removes more hair. ______________

14. Explain how to hold thinning shears. ______________________

15. Describe how to hold the comb and scissors during the haircutting process.

____

____

____

16. Name a time-saving rule to observe while cutting hair.

____

HAIR THINNING

17. Explain the purpose of thinning hair.

____

18. Explain why fine hair can be thinned closer to the scalp than coarse hair.

____

____

19. Specify the distances from the scalp for thinning the following hair textures:

    Fine hair: ______________

    Medium hair: ______________

    Coarse hair: ______________

20. List four areas where hair thinning is not advisable.

    1. ______________________ 2. ______________________

    3. ______________________ 4. ______________________

21. Explain why hair should not be thinned near the ends of a strand.

____

22. Explain why cosmetologists should be cautious when thinning hair.

____

____

23. Describe how to grip hair firmly and evenly.

____

24. Specify the size of the sections to use when thinning hair.

____

25. Explain how to hold hair sections for thinning.

____

26. Specify how far apart each cut should be made on the hair strand. ______________

27. Define slithering or effilating. ______________________________

28. Describe what to do with the scissors during slithering.

____

____

HAIRCUTTING WITH SCISSORS

29. Describe what is done to hair before a dry shaping.

____

30. Describe what is done to hair before a wet shaping.

____

31. List three client characteristics that must be analyzed before beginning the haircut.

    1. ______________________ 2. ______________________

    3. ______________________

32. Specify who decides on a suitable haircut before beginning. ______________________

33. Define nape guideline hair.

____

34. List the order in which the guideline strands are cut.

____

35. List the order in which the rest of the hair is cut after the guideline.

____

36. Explain how to hold hair sections throughout the haircut. ______________________

37. Describe how to determine the correct length of strands to be cut as the haircut progresses.

____

38. Specify where the cosmetologist works for even cutting of bangs.

____

39. Identify what to use as a guide when cutting short bangs. ______________________

40. List the two steps necessary to complete the haircut before proceeding to the next professional service.

    1. ____

    2. ____

SHINGLING

41. Define shingling.

____

____

42. Specify how to position the client's head for shingling. ______________________

43. Identify the area where shingling is started. ______________________

44. Describe how to hold the scissors and comb for shingling.

____

45. Describe the action of the top blade during shingling.

____

USE OF CLIPPERS ON THE NECKLINE

46. Explain why using clippers on the neckline cannot make hair grow in thicker on the neck.

____

____

USING THE RAZOR

47. Describe the finger-wrap hold.

____

____

48. Describe the position of the guard while working. ______________________

49. Describe the three-finger hold.

____

____

____

50. Identify in which hand to hold the razor and comb when combing the hair.

____

51. Identify in which hand to hold the razor and comb when cutting the hair.

____

52. Give two reasons for keeping hair damp when cutting with a razor.

    1. ____________________ 2. ____________________

53. Specify which part of the guard must be over the cutting edge of the blade.

____

54. Explain how to place the razor on the hair for thinning.

____

55. Describe the strokes used to thin hair with a razor. ______________________

56. Describe where the razor strokes are directed. ______________________

57. List the order in which guideline strands are cut.

____

58. List the order in which the rest of the hair is cut after the guideline.

____

59. Describe how to determine the correct length of strands to be cut as the haircut progresses.

____

60. List the two steps necessary to complete the haircut before proceeding to the next professional service.

    1. ____

    2. ____

LEARN TO HANDLE CHILDREN

61. Identify a characteristic necessary to attract children and their parents to the salon.

____

CUTTING OVERLY CURLY HAIR

62. Identify at what stage in the haircutting process overly curly hair is shampooed and dried.

____

63. Explain how to comb overly curly hair before cutting.

____

____

____

64. Describe how to begin cutting overly curly hair. _______________________

65. Explain how to continue cutting overly curly hair.

____

____

____

66. Name two implements used to outline overly curly hair. ___________________

WORD REVIEW

blunt
crown
finger-wrap hold
guideline
nape
parting
sectioning
shaping
slithering
three-finger hold
bulk
effilating
guard
haircutting
notched
pliable
serrated
shingling
texturizing
trimmer
clippers
emollient
guide
hairstyles
parallel
section
shank
silhouette
thinning

MATCHING TEST

Insert the correct term or phrase in front of each definition.

effilating
shaping
thinning
guideline
shingling
sectioning
slithering

1. ________________ thinning hair with scissors
2. ________________ cutting the hair close to the nape and gradually longer toward the crown
3. ________________ the section of hair that serves as a length guide
4. ________________ removing excess bulk from hair without shortening its length
5. ________________ another term for slithering

RAPID REVIEW TEST

Place the correct word in the space provided in each sentence below.

accentuate
clippings
follicles
instructor
parallel
ring finger
single
texture
blades
elasticity
foundation
length
perpendicular
scalp
splits
thumb
check
ends
guard
minimize
practice
shape

1. In shingling, the blades of the scissors are held ______________ with the comb.
2. A good haircut serves as a ________________ for attractive hairstyles and for other services performed in the salon.
3. During slithering, the scissors are closed slightly each time they are moved toward the ____________.
4. Before cutting overly-curly hair, it must be combed so that its _____________ is extended as much as possible.
5. When combing hair during a cut, it's important to close the _____________ of the shears and remove the _____________ from the ring.
6. When selecting a suitable hairstyle, the client's head _____________, facial contour, neckline, and hair _____________ must be considered.
7. When razor cutting, the ____________ must face the cosmetologist.
8. It's important for cosmetologists to ____________ all hair cuts for proper length.
9. Hairstyles should ________________ the client's good points and _______________ the poor features.
10. Of the two types of thinning shears, the ____________-notched cuts more hair.
11. Instruction in haircutting must be followed by continual ______________ under the guidance of an ________________.
12. The direction of razor strokes on the hair is toward the ____________.
13. Combing overly-curly hair in a circular pattern before cutting helps to avoid __________.
14. Completing a haircut involves the thorough cleaning of all hair _______________ from the cape, client's clothing, and the work area.
15. The use of hair clippers on the neckline does not increase the number of hair ___________.

MULTIPLE CHOICE TEST

Read each statement carefully, then write the letter representing the word or phrase that correctly completes the statement on the blank line to the right.

1. The still blade of the haircutting scissors is controlled by the

   a) thumb   b) middle finger
   c) ring finger   d) little finger   ________

2. Double-notched shears are used to

   a) shorten hair   b) outline hair
   c) thin hair   d) blunt-cut hair   ________

3. Hair may be thinned on/at the

   a) top   b) hair part
   c) facial hairline   d) hair ends   ________

4. Cutting hair close to the nape and gradually longer toward the crown is known as

   a) thinning   b) slithering
   c) effilating   d) shingling   ________

5. Hair is held firmly and evenly by overlapping the middle finger slightly over the

   a) thumb   b) index finger
   c) ring finger   d) little finger   ________

6. To thin hair with a razor, it should be held

   a) at a 30° angle   b) at a 45° angle
   c) flat against the strand   d) erect against the strand   ________

7. Once hair is cut too short, it may be difficult to achieve the desired

   a) hairstyle   b) hair color
   c) hair texture   d) hair elasticity   ________

8. A guide to use when cutting short bangs is the

   a) top of head   b) crown of head
   c) earlobe   d) bridge of nose   ________

9. Overly-curly hair is outlined at the hairline with scissors or

   a) straight razor   b) razor with safety guard
   c) thinning shears   d) trimmer   ________

10. The hair that can be thinned closest to the scalp is

    a) fine hair   b) medium hair
    c) coarse hair   d) curly hair   ________

11. The strokes needed to thin hair with a razor are

    a) short and jerky   b) short and steady
    c) long and steady   d) long and jerky   ________

12. Hair is shampooed and completely dried before a

    a) hair color   b) hair set
    c) wet shaping   d) dry shaping   ________

13. A characteristic necessary for handling children in the salon is

a) patience
b) moodiness
c) irritability
d) bitterness
____

14. The four-section parting involves parting the hair from forehead to

a) crown
b) top
c) nape
d) sides
____

15. Thinning near the ends of a strand causes the hair to be

a) curly
b) straight
c) colorless
d) shapeless
____

16. Hair must be damp when it is shaped with

a) scissors
b) razor
c) hair clippers
d) thinning shears
____

17. Hair that must be thinned furthest from the scalp is

a) fine hair
b) medium hair
c) coarse hair
d) curly hair
____

18. The moving blade of haircutting scissors is controlled by the

a) thumb
b) middle finger
c) ring finger
d) little finger
____

19. Hair may be thinned with razor, thinning shears, or

a) clipper
b) outliner
c) trimmer
d) scissors
____

20. To cut bangs evenly, the cosmetologist must work directly

a) in back of client
b) in front of client
c) to right side of client
d) to left side of client
____

Date ______________

Rating ______________

Text Pages 101–110

FINGER WAVING

INTRODUCTION

1. Define finger waving.

____

____

____

2. Give three reasons why training in finger waving is important.

   1. ____
   2. ____

      ____
   3. ____

      ____

PREPARATION

3. Explain how to prepare the client for finger waving.

____

____

4. Describe the type of hair most suitable for finger waving.

____

FINGER WAVING LOTION

5. Name two reasons for using waving lotion.

   1. ____
   2. ____

6. List two qualities of a good waving lotion.

   1. ______________________ 2. ______________________

APPLICATION OF LOTION

7. Explain how to make hair move more easily. ______________________________

8. Explain why the natural growth must be followed when combing and parting the hair.

____

9. Explain why waving lotion is applied to hair while it is damp.

____

10. Describe what to avoid when applying waving lotion.

____

11. Explain how to determine the natural hair growth.

____

____

HORIZONTAL FINGER WAVING

12. Describe how to shape the top area.

____

____

13. Explain how the comb and index finger are positioned to begin forming the first ridge.

____

____

____

14. Describe how to hold the first ridge in place.

____

15. Describe how to emphasize ridges.

____

16. Explain what happens when ridges are pinched or pushed with the fingers.

____

17. Name what should not be visible if ridges and waves are matched evenly.

____

18. Specify the direction of the movements required for forming the second ridge.

____

19. Explain the direction of the movements required for forming the third ridge.

____

____

20. Describe how the client's hair is secured before drying.

____

21. Describe how to protect client's forehead and ears while under the dryer.

22. List the items that must be sanitized after each use.

ALTERNATE METHOD OF FINGER WAVING

23. Differentiate between the alternate method of finger waving and the horizontal finger wave.

VERTICAL FINGER WAVING

24. Differentiate between vertical and horizontal finger waves.

25. Describe the part required for vertical finger waving.

26. Describe the effect of the shaping required for vertical finger waving.

27. Identify which of the illustrated finger waves is horizontal and which is vertical.

SHADOW WAVE

28. Define shadow wave. ______________________________

29. Explain the technique required to make shadow waves.

____

30. Explain when a shadow wave is desirable.

____

REMINDERS AND HINTS ON FINGER WAVING

31. Describe the type of comb used for finger waving.

____

32. Name what should be located before starting the finger waving.

____

33. Explain how to wave underneath hair. ______________________________

34. Explain how to achieve longer-lasting waves.

____

35. Describe the effects of prolonged drying.

____

36. Explain what happens when hair is not thoroughly dried before combing out.

____

37. Describe how to handle hair that tangles easily. ______________________________

38. List two benefits of spraying hair with lacquer.

    1. ______________________________ 2. ______________________________

WORD REVIEW

dexterity
growth pattern
ridge
waving lotion
emphasize
horizontal
shadow wave
finger wave
pliable
vertical

RAPID REVIEW TEST

Place the correct word in the space provided in each sentence below.

alternate	circular	coarse
excessive	facial	fine
flake	hands	horizontal
over	pressure	protectors
reverse	same	scalp
separations	shadow	under
vertical		

1. A good finger waving lotion does not ____________ when it dries.
2. Ridges are emphasized by closing the middle and index fingers and applying ______________ to the head.
3. The movements necessary to form the third ridge are the ____________ as those followed in forming the first ridge.
4. Combs used for finger waving should have both ___________ and _____________ teeth.
5. Cosmetologists should wash their _____________ before beginning a finger wave.
6. Before forming the first ridge, hair is shaped in a ______________ movement.
7. The ridges and waves run up and down the head in ______________ finger waving.
8. To wave the underneath hair, the comb must be inserted through the hair to the _________.
9. A finger wave correctly done complements the client's head as well as her ____________ features.
10. The ridge and wave of each section should match evenly, without showing _________________ in the ridge and hollow part of the wave.
11. Pinching with fingers to try to increase the height or depth of ridges creates ____________-direction of the ridges.
12. The ridges and waves run parallel around the head in ________________ finger waving.
13. Cotton, gauze, or paper ________________ are used to safeguard the client's forehead and ears while under the dryer.
14. The movements necessary to form the second ridge are the _____________ of those followed in forming the first ridge.
15. It's important to avoid applying an _______________ amount of waving lotion to the hair.

MULTIPLE CHOICE TEST

Read each statement carefully, then write the letter representing the word or phrase that correctly completes the statement on the blank line to the right.

1. A good finger waving lotion
   a) dries slowly
   b) colors the hair
   c) makes hair sticky
   d) is harmless to the hair ________

2. Before placing client under dryer, the finger wave is protected by a
   a) turban
   b) plastic bag
   c) hairnet
   d) towel ________

3. The best results in finger waving are obtained when the hair is
   a) straight
   b) naturally curly
   c) frizzy
   d) overly curly ________

4. Finger waving is the art of shaping and directing the hair into waves using fingers and
   a) rollers
   b) waving rods
   c) comb
   d) curling irons ________

5. Finger waves will hold longer if sprayed lightly with
   a) hair color
   b) cream rinse
   c) waving lotion
   d) lacquer ________

6. A shadow wave is a wave with
   a) high ridges
   b) deep ridges
   c) sharp ridges
   d) low ridges ________

7. The natural oils of the hair and scalp will dry out when subjected to
   a) prolonged drying
   b) shampooing
   c) rinsing
   d) conditioning ________

8. Combs used for finger waving should be made of
   a) soft rubber
   b) hard rubber
   c) plastic
   d) metal ________

9. Natural hair growth is determined by combing hair away from the face, then pushing it
   a) forward
   b) backward
   c) toward the right side
   d) toward the left side ________

10. Lightened or tinted hair that tangles is easier to comb after applying
    a) hair color
    b) cream rinse
    c) waving lotion
    d) lacquer ________

11. When forming ridges, the comb is inserted directly
    a) under the middle finger
    b) over the middle finger
    c) under the index finger
    d) over the index finger ________

12. Waving lotion is applied to the hair while it is
    a) damp
    b) dry
    c) dripping wet
    d) thoroughly wet ________

13. Finger waving provides valuable training in creating

a) hairstyles
b) haircuts
c) hair colors
d) hair relaxers ________

14. Combs, hairpins, clippies, and hairnet are sanitized

a) after each use
b) once a week
c) once a month
d) once a year ________

15. Before beginning a finger wave, it's important to locate the

a) natural color
b) natural wave
c) receding hairline
d) amount of gray hair ________

16. Hair is kept in place during finger waving by

a) hair color
b) cream rinse
c) waving lotion
d) lacquer ________

17. Finger waves will not last if hair is combed out before it is thoroughly

a) shampooed
b) conditioned
c) relaxed
d) dried ________

18. Shadow waves are desirable for clients who wear their hair

a) high on the top
b) high in the crown
c) full at the nape
d) close to the head ________

19. Finger waving helps cosmetologists develop dexterity, coordination, and

a) back strength
b) finger strength
c) good posture
d) long fingernails ________

20. Lightly spraying hair with lacquer adds

a) color
b) curl
c) sheen
d) wave ________

Date ______________

Rating ______________

Text Pages 111–162

WET HAIRSTYLING

INTRODUCTION

1. List eight areas that cosmetologists must first understand to become proficient stylists.

   1. ______________ 2. ______________
   3. ______________ 4. ______________
   5. ______________ 6. ______________
   7. ______________ 8. ______________

2. List ten elementary rules of art used to style hair.

   1. ______________ 2. ______________
   3. ______________ 4. ______________
   5. ______________ 6. ______________
   7. ______________ 8. ______________
   9. ______________ 10. ______________

3. Explain why an examination of the client's hair is necessary before starting the shampoo.

   ____

   ____

REMOVING TANGLES FROM HAIR

4. Explain why removing tangles from hair is necessary before shampooing, cutting, or styling. ______________________________

5. Identify the area on the head at which to begin removing tangles. ______________

6. Name two implements used to remove tangles.

   1. ______________ 2. ______________

7. Specify the direction in which hair is combed to remove tangles.

   ____

8. List five characteristics that determine size of hair sections.

   1. ______________ 2. ______________
   3. ______________ 4. ______________
   5. ______________

MAKING A PART

9. Specify the direction in which hair is combed to make a straight part.

____

10. Describe how to finish parting the hair.

____

____

____

11. Describe how to find the natural part.

____

____

PIN CURLS

12. List three types of hair that can be pin curled.

    1. ____________________ 2. ____________________

    3. ____________________

13. Describe when pin curls work best.

____

14. Name the parts of a curl.

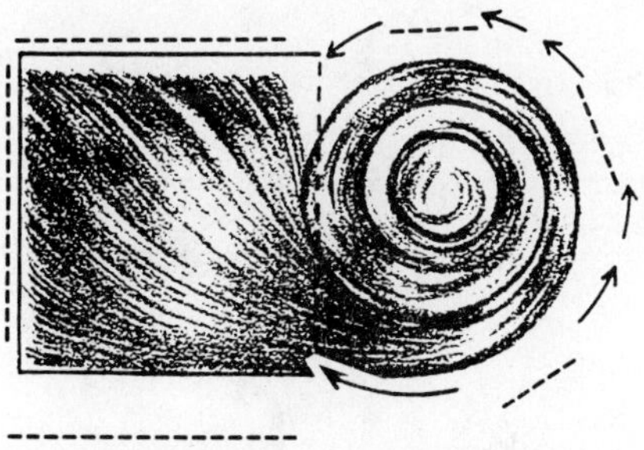

15. Define base.

____

16. Define stem.

____

____

17. Define circle.

____

____

18. Define mobility. ____________________________________

19. Specify what determines the mobility of a section of hair. ______________________

20. Explain how to form the no-stem curl. ______________________

21. Describe the results obtained from the no-stem curl. ______________________

22. Explain how to form the half-stem curl.

____

23. Describe the results obtained from the half-stem curl.

____

24. Explain how to form the full-stem curl. ______________________

25. Describe the results obtained from the full-stem curl.

____

26. Describe the results obtained from open-center curls.

____

27. Describe the results obtained from closed-center curls.

____

28. Identify the type of curl recommended for fine hair or a desired fluffy curl.

____

29. Explain what determines the size of the wave. ______________________

30. Identify which of the illustrated curls is an open-center curl and which is a closed-center curl.

______________________ ______________________

31. Explain what determines the direction of the finished curl.

____

32. Identify the movement of the forward curl. ______________________

33. Identify the movement of the reverse curl. ______________________

34. Label each illustration for stem and curl directions.

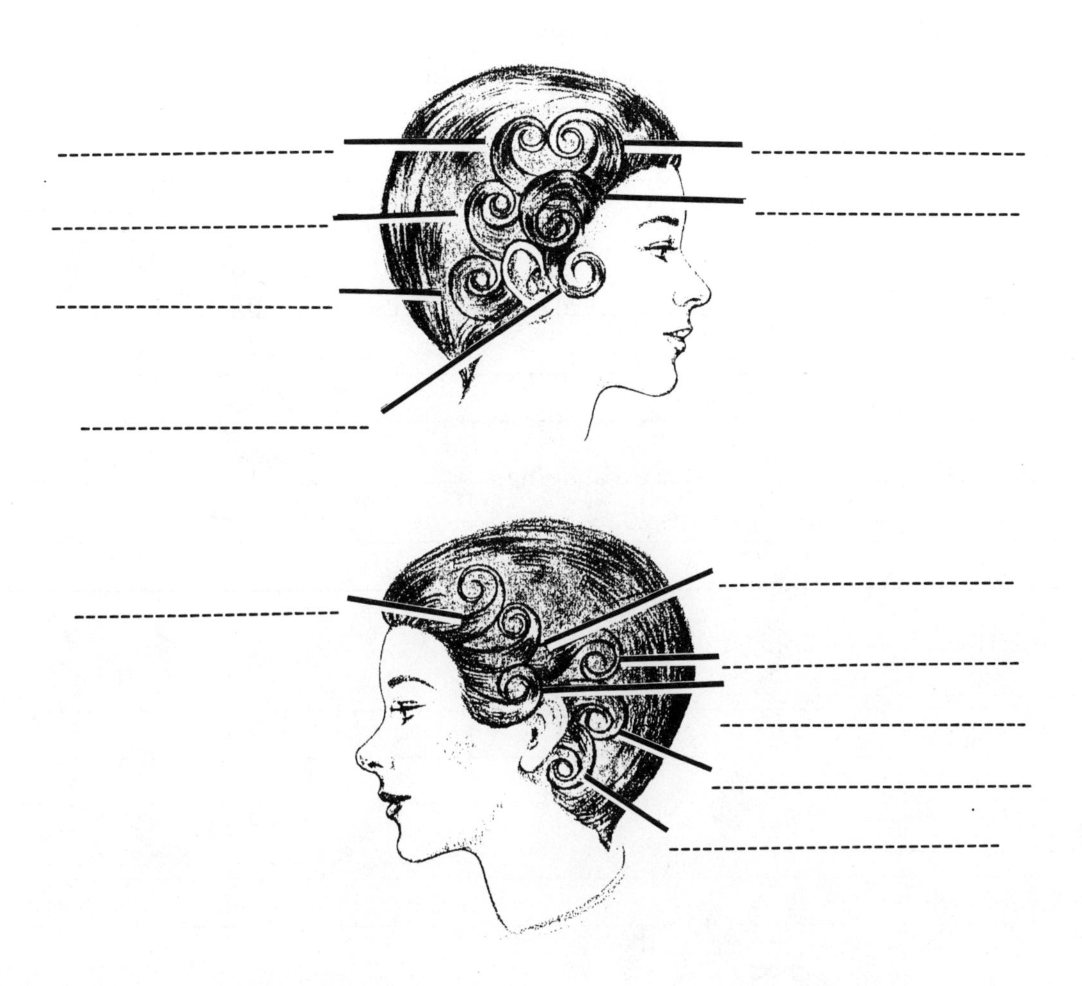

35. Describe clockwise curls.

____

36. Describe counterclockwise curls.

____

37. Identify which of the illustrated curls are clockwise and which are counter-clockwise.

______________________ ______________________

38. Define shaping.

39. Describe circular shapings.

40. Describe oblong shapings.

41. Describe forward shapings.

42. Explain how to make a forward, vertical shaping on the side of the head.

43. Explain how to make a top forward shaping.

44. Describe reverse shapings.

45. Describe diagonal shapings.

46. Describe vertical side shapings.

47. Describe horizontal shapings.

48. Explain under what conditions horizontal shapings are recommended.

PIN CURL FOUNDATIONS OR BASES

49. List the four most commonly shaped bases.

1.
2.
3.
4.

50. Explain why stylists must use care when selecting and forming the base curl.

51. Explain how to achieve further uniformity of curl development.

____

52. Describe how each curl must lie. ____

53. Explain where and why rectangular base pin curls are recommended.

____

54. Name the reason that pin curls must overlap. ____

55. Explain where and why triangular base pin curls are recommended.

____

56. Specify how arc base (half-moon or C-shape) pin curls are formed.

____

57. Describe the results obtained from arc base pin curls.

____

____

58. Explain why square base pin curls are used.

____

59. Explain how to avoid splits when combing out square base pin curls.

____

PIN CURL TECHNIQUES

60. Describe carved curls.

____

61. Label the open and closed ends of a shaping.

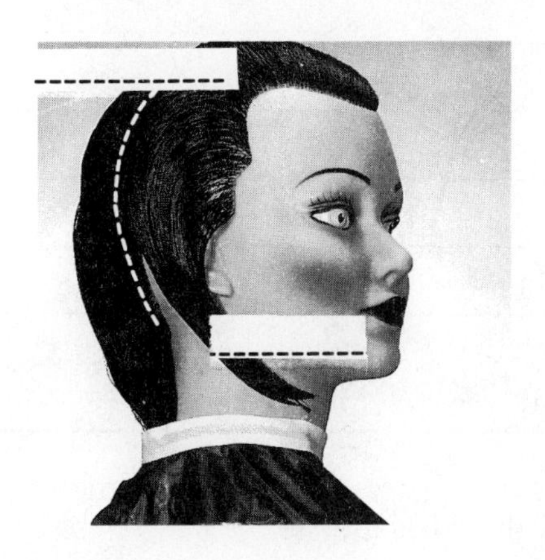

62. List the eleven steps required for forming pin curls on the right side.

1. ______
2. ______
3. ______
4. ______
5. ______
______
6. ______
7. ______
8. ______
9. ______
10. ______
11. ______

63. Identify what should not be destroyed while combing or pinning the curl.

______

64. Explain how to achieve longer-lasting curl movement.

______
______

65. List the eight steps required for forming pin curls on the left side.

1. ______
______
2. ______
3. ______
4. ______
5. ______
6. ______
7. ______
8. ______

66. Specify within what pin curls must fit. ______

67. Describe the graduation of the size of the curls.

______

68. Describe the position of the comb while slicing a strand.

____

69. Explain which teeth of the comb to use to ribbon hair.

____

70. Describe how to ribbon the tip of the strand.

____

ANCHORING PIN CURLS

71. Explain the importance of correctly anchoring pin curls.

____

____

72. Describe how clips are always inserted. _______________________

73. Describe how to anchor the pin curl correctly.

____

____

74. Explain how to avoid indentations or impressions across the hair.

____

75. Explain why clippies or hairpins are desirable for securing curls made with small strands or fine hair.

____

76. Explain why clippies or hairpins are not desirable for securing curls of coarse or thick hair.

____

77. Describe how to protect clients from clips that touch the skin during the drying process.

____

78. Identify the type of curl (clockwise or counter-clockwise) and the shaping (forward or reverse) for each of the illustrated pincurl patterns.

_______________ _______________

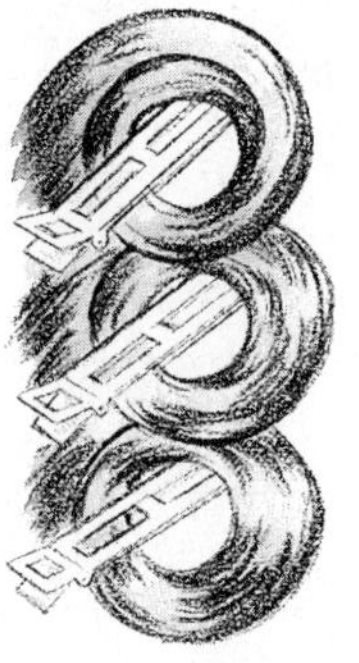

_______________ _______________

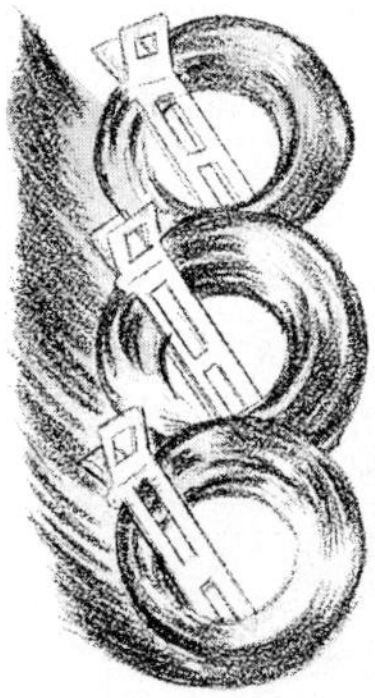

_______________ _______________

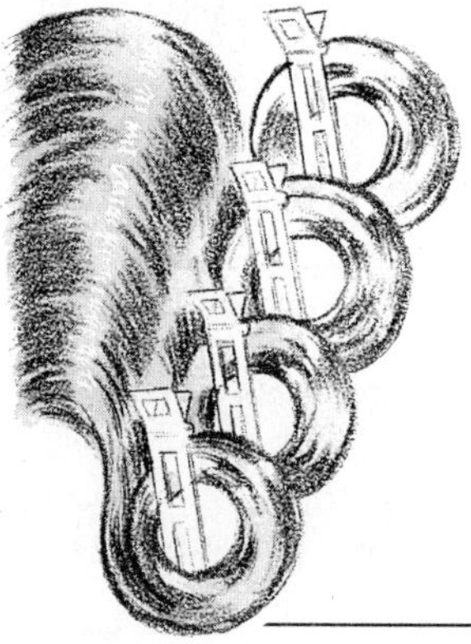

EFFECTS OF PIN CURLS

79. Explain how to avoid fighting uneven wave or curl design during the comb-out.

____

____

80. Describe how to begin a vertical wave for the best result.

____

81. Describe how to shape a horizontal wave.

____

82. Explain how to produce clash in an interlocking movement.

____

____

83. Identify the direction in which to shape hair for a diagonal wave. _______________

84. Explain how to wave bangs.

____

85. Describe how to set a French twist.

____

____

86. Describe how to comb out a French twist.

____

____

____

87. Explain when to use ridge curls. ______________________

88. Describe how to form skip waves.

____

____

89. Draw a line from each pin curl pattern to the correct comb-out.

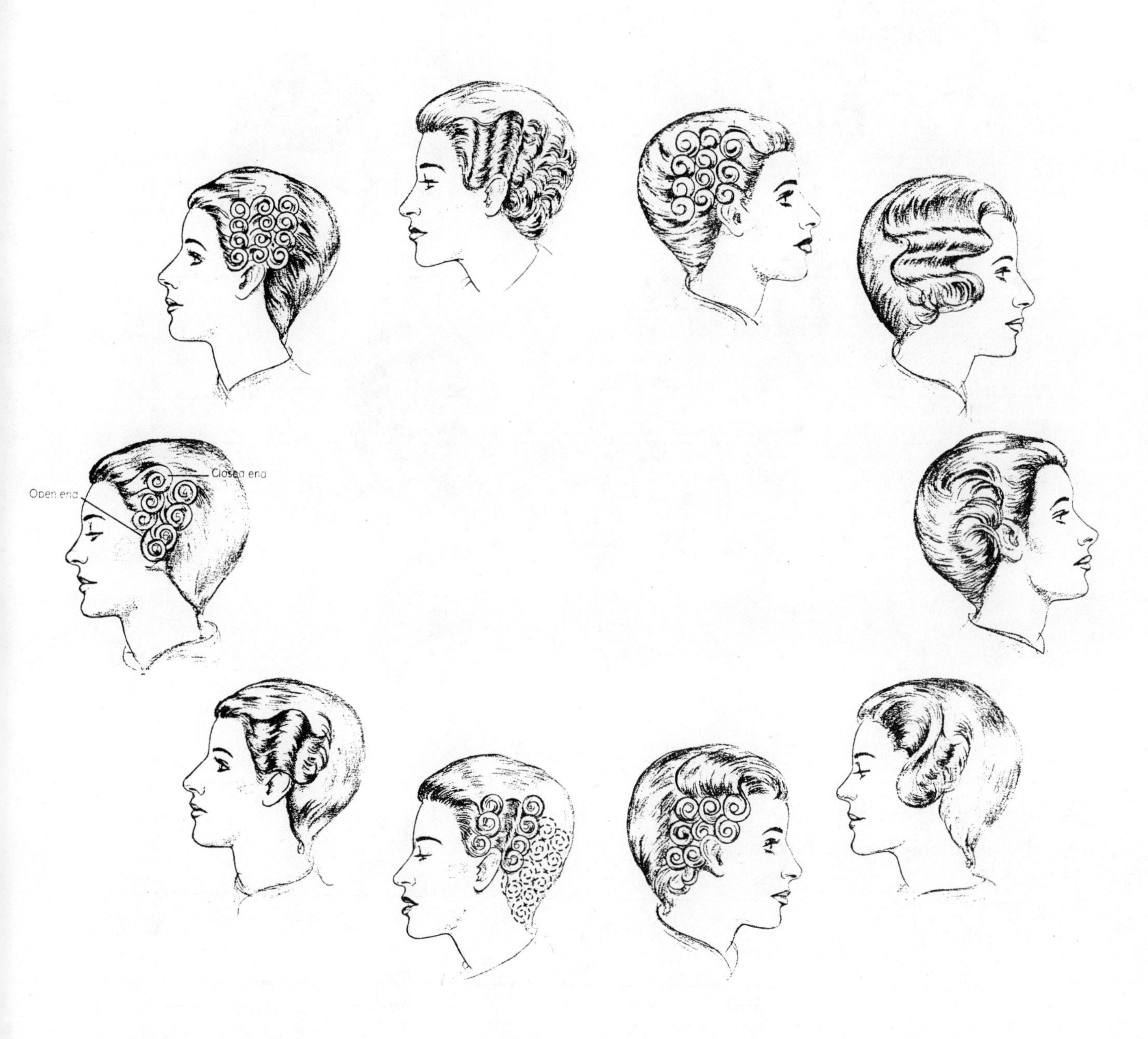

90. Draw a line from each pin curl pattern to the correct comb-out.

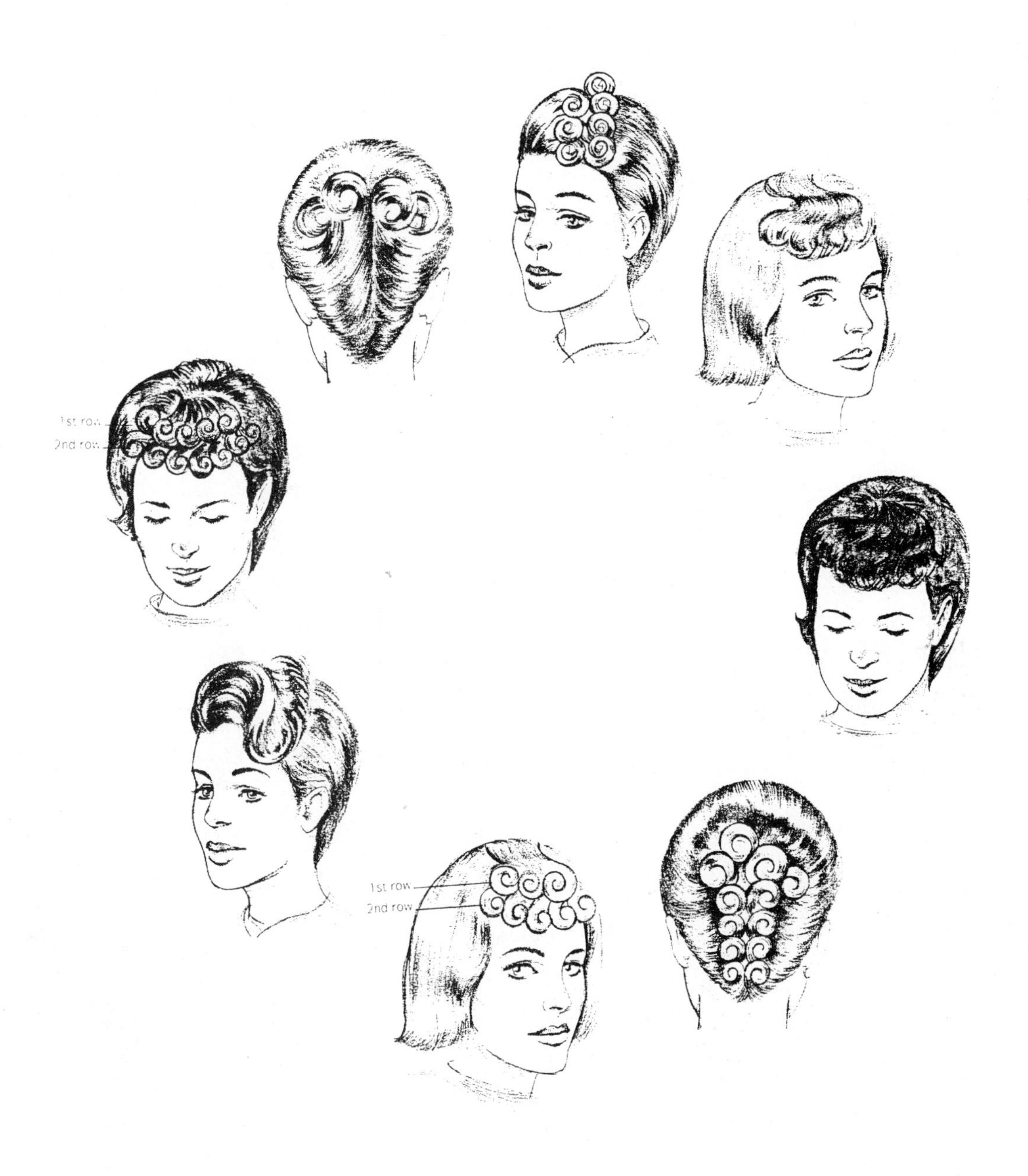

THE CASCADE OR STAND UP CURL

91. Explain when to use the cascade or stand up curl.

____

92. Describe how to form a cascade or stand up curl.

____

____

SEMI-STAND UP CURLS

93. Describe how to carve and pin semi-stand up curls.

____

94. Name the effect that can be achieved with semi-stand up curls. _______________

ROLLER CURLS

95. Name two effects created by rollers.

    1. _______________ 2. _______________

96. List four advantages to using rollers instead of stand up curls.

    1. _______________
    2. _______________
    3. _______________
    4. _______________

97. Specify the size of the base needed for a roller 3" long and 1" wide.

____

98. Explain why hair must be fully saturated.

____

99. Describe how to control hair ends when using rollers.

____

____

100. Identify the length of hair needed to wind around rollers.

____

101. Compare winding a strand of hair 4 1/2" long around rollers of various diameters.

1": ____

____

3/4": ____

1/2": ____

102. Draw a line from each pin curl pattern to the correct comb-out.

BARREL CURL

103. Explain when to use a barrel curl.

____

104. Describe how to make a barrel curl.

____

105. Explain how to create volume.

____

106. Draw a line from the illustrated angle of hair strand to the correct description of its effect.

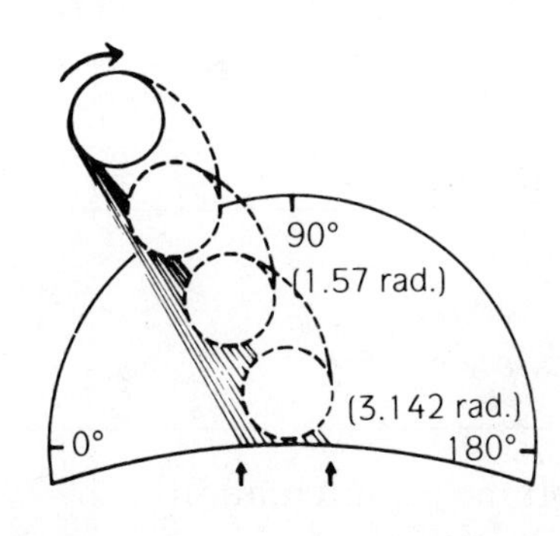

Less volume

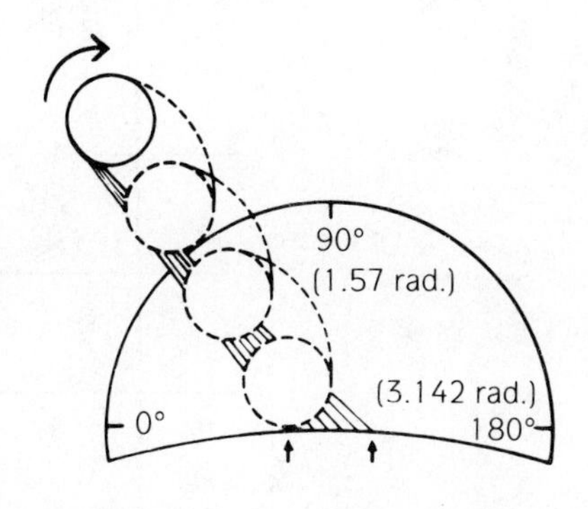

Full volume

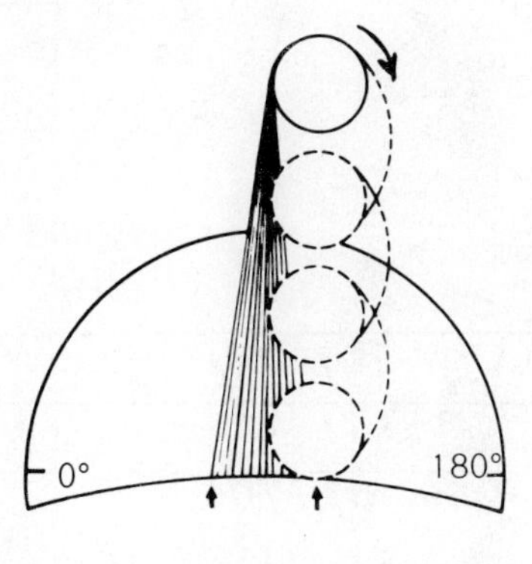

Indentation

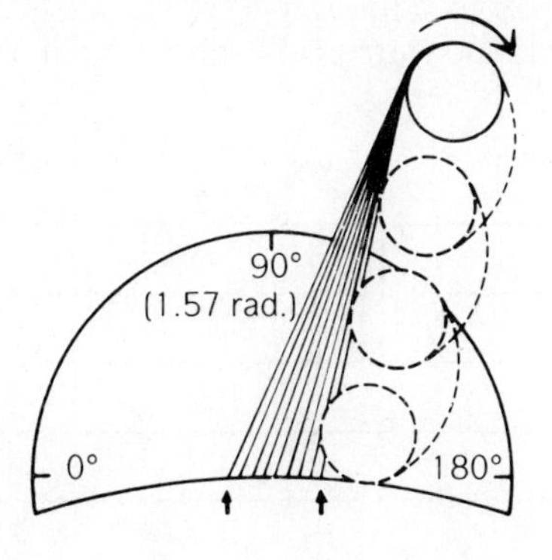

Medium volume

Added volume

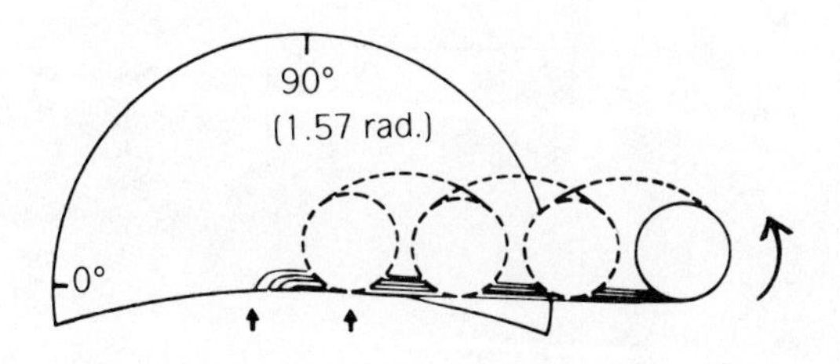

107. Describe where the roller sits for full volume. ______________________________

108. Describe where the roller sits for medium volume.

____

109. Describe where the roller sits for a small amount of lift. ______________________

110. Describe how to create indentation or hollowness.

____

111. Indicate which of the illustrated rollers produces volume or indentation.

1- ______________

2- ______________

3- ______________

4- ______________

5- ______________

112. List other names for circular roller action.

1. ______________________ 2. ______________________
3. ______________________ 4. ______________________
5. ______________________ 6. ______________________
7. ______________________

113. List ten other names for the spot or area from which the hair is directed to form a circular movement.

1. ______________________ 2. ______________________
3. ______________________ 4. ______________________
5. ______________________ 6. ______________________
7. ______________________ 8. ______________________
9. ______________________ 10. ______________________

114. Draw a line from each set to the correct comb-out.

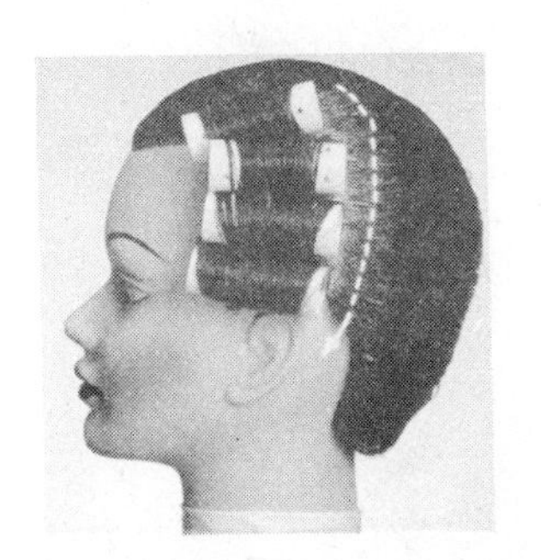

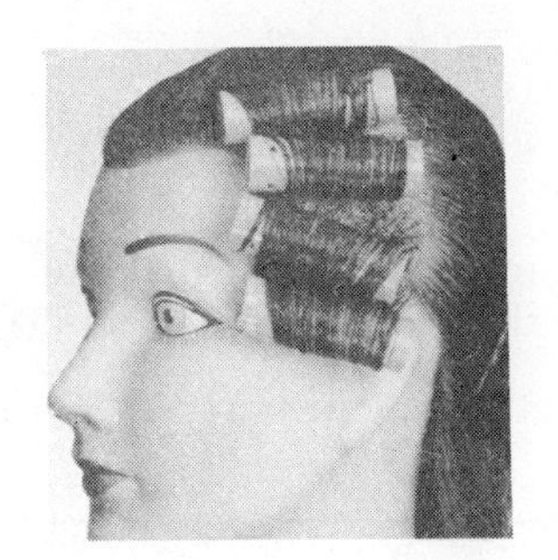

115. Draw a line from each set to the correct comb-out.

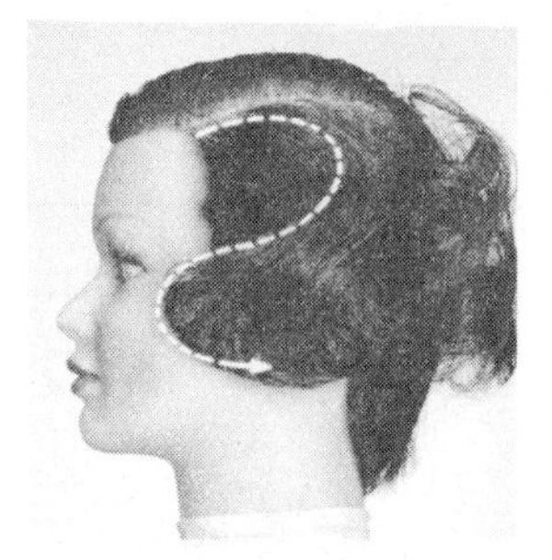

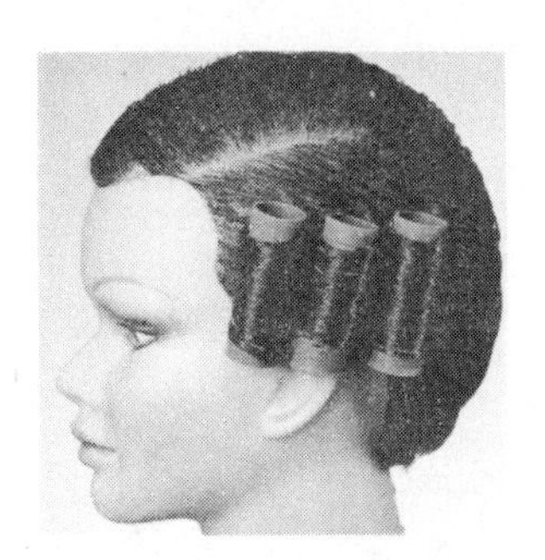

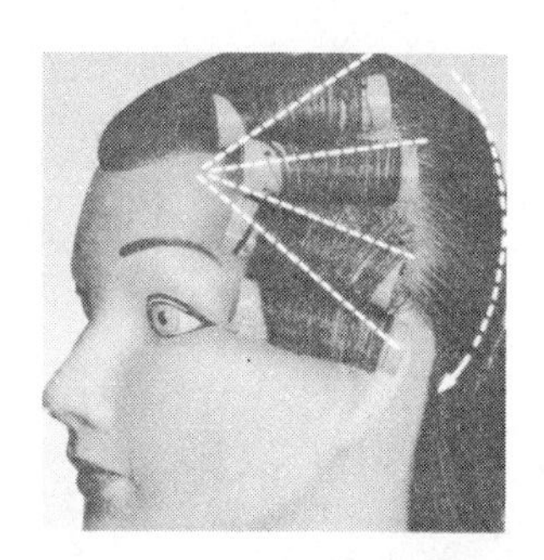

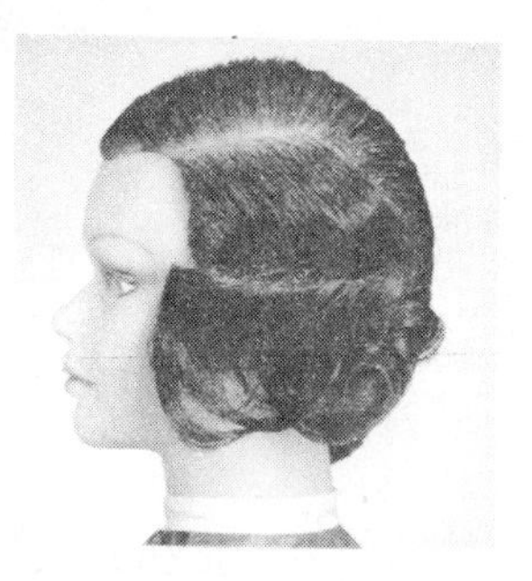

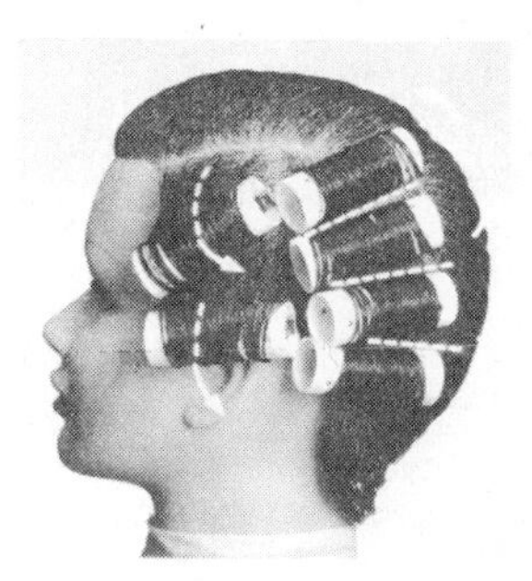

116. Explain why the circular movement of the hair might be weaker when using cylinder-shaped instead of tapered rollers.

____

____

117. Compare the sizes of rollers needed for fine and for coarse hair.

____

COMB-OUT TECHNIQUES

118. Identify what precedes smooth and well-executed comb-outs. ____________________

119. List two reasons for brushing the hair after removing rollers and clips.

1. ____________________ 2. ____________________

120. Explain why hair is smoothed and brushed into a semi-flat condition.

____

121. Describe how to direct hair into the general pattern desired.

____

____

122. Explain why lines of direction should be slightly overemphasized.

____

123. Explain which areas need back-combing and which need back-brushing.

____

124. List three qualities that should be revealed in finished patterns.

1. ______________ 2. ______________

3. ______________

125. Explain why final touches are important. ______________

126. Explain why the tail of a comb is used.

____

127. Name the two final steps in the comb-out.

1. ______________ 2. ______________

128. Describe how to back-comb and back-brush hair.

____

____

129. List four other names for back-combing.

1. ______________ 2. ______________

3. ______________ 4. ______________

130. Explain when to use back-combing.

____

131. Specify the width of sections for back-combing. ______________

132. Describe how hair is held for back-combing. ______________

133. Name the implement that is used for back-combing. ______________

134. Specify the distance from the base to insert comb. ______________

135. Describe the action of the comb to finish back-combing.

____

____

____

____

136. Give another name for back-brushing. ______________

137. Explain when to use back-brushing.

____

____

138. Describe how to hold hair for back-brushing.

____

139. Name the implement that is used for back-brushing. _______________

140. Describe where to insert the brush. _______________

141. Describe the action of the brush to finish back-brushing.

____

____

____

____

____

BRAIDING

142. Describe how braids must be done. _______________

143. Explain why braiding on damp hair is advisable.

____

144. Name two types of French braids.

1. _______________ 2. _______________

145. Describe how invisible braiding is done. _______________

146. List the seven steps required to form an invisible braid.

1. _______________
2. _______________
3. _______________
4. _______________
5. _______________

____

____

6. _______________

____

7. _______________

____

147. Describe how to keep the braid neat with all short ends in place.

____

148. Describe two variations for braiding long hair.

1. ______

2. ______

149. Describe three variations for braiding hair that is not very long.

1. ______

2. ______

3. ______

150. Describe how visible braiding is done. ______

151. Explain how the hair is parted and sectioned for visible braiding.

______

152. List the five steps required to form a visible braid.

1. ______

______

______

2. ______

______

______

3. ______

______

4. ______

5. ______

153. Explain how corn-rowing is done.

______

______

154. Identify the type of hair on which corn-rowing is recommended. ______

155. Specify the length of time that corn-rowing should last. ______

156. List the five steps necessary to prepare curly hair for corn-rowing.

1. ______

2. ______

3. ______

4. ______

5. ______

______

ARTISTRY IN HAIRSTYLING

157. Describe what the cosmetologist's job is in designing attractive hairstyles.

____

____

158. List three qualities that each client deserves in a hairstyle.

    1. _______________
    2. _______________
    3. _______________

159. List three general characteristics upon which artistic and suitable hairstyles are based.

    1. _______________ 2. _______________
    3. _______________

FACIAL TYPES

160. Name what determines facial shapes. _______________

161. List the seven facial shapes.

    1. _______________ 2. _______________
    3. _______________ 4. _______________
    5. _______________ 6. _______________
    7. _______________

162. List the three facial zones.

    1. _______________ 2. _______________
    3. _______________

163. Name the face shape that is considered the ideal shape. _______________

164. Describe the contour of the oval facial type.

____

____

165. Identify the type of hairstyle best suited for the client with an oval-shaped face.

____

166. Describe the contour of the round facial type.

____

167. Explain how to style the hair for a client with a round facial type.

____

____

____

168. Describe the contour of the square facial type.

____

169. Explain how to style the hair for a client with a square facial type.

____

____

170. Describe the contour of the pear-shaped facial type.

____

171. Explain how to style the hair for a client with a pear-shaped facial type.

____

____

172. Describe the contour of the oblong facial type.

____

173. Explain how to style the hair for a client with an oblong facial type.

____

____

174. Describe the contour of the diamond facial type.

____

175. Explain how to style the hair for a client with a diamond-shaped facial type.

____

____

176. Describe the contour of the heart-shaped facial type.

____

177. Explain how to style the hair for a client with a heart-shaped facial type.

____

____

178. Name the profile that is considered the ideal. _______________

179. Describe the straight profile.

____

180. Describe how to style the hair for a client with a concave profile.

____

181. Describe how to style the hair for a client with a convex profile.

____

182. Describe how to style the hair for a client with a low forehead and protruding chin.

____

____

____

183. Name two views of the nose that must be considered.

1. ______________________ 2. ______________________

184. Describe the turned-up nose. ______________________

185. Identify the type of hairstyle to avoid when styling the hair of a client with a turned-up nose. ______________________

186. Describe how to style the hair for a client with a turned-up nose.

____

____

187. List three types of nose shapes classified as prominent.

1. ______________________ 2. ______________________

3. ______________________

188. Explain the cosmetologist's objective when styling the hair of a client with a prominent nose. ______________________

189. Describe how to style the hair for a client with a prominent nose.

____

190. Describe how to style the hair for a client with a crooked nose.

____

191. Explain why a well-balanced hairstyle should be avoided on a client with a crooked nose.

____

192. Identify the problem presented by a wide, flat nose. ______________________

193. Describe how to style the hair for a client with a wide, flat nose.

____

194. Explain the relationship of the eyes to the face. ______________________

195. Name two face shapes associated with wide-set eyes.

1. ______________________ 2. ______________________

196. Explain how hair is styled to minimize the effect of wide-set eyes.

____

____

197. Name the face shape associated with close-set eyes. ______________________

198. Explain the cosmetologist's objective when styling the hair for a client with close-set eyes.

____

199. Describe how to style the hair for a client with close-set eyes.

____

____

200. Name the ideal head shape. ______________________________

201. Identify the cosmetologist's goal when designing hairstyles.

____

202. Explain how to deal with flat areas on the head. ______________________________

203. Explain the purpose of analyzing clients' features. ______________________________

204. Identify the aim when styling the hair for a client with a short, plump neck.

____

205. Describe how to style the hair for a client with a short, plump neck.

____

206. Identify the aim when styling the hair for a client with a long, thin neck.

____

207. Describe how to style the hair for a client with a long, thin neck.

____

208. Identify the aim when styling the hair for a client with thin features.

____

209. Describe how to style the hair for a client with thin features.

____

____

210. Identify the aim when styling the hair for a client with uneven features.

____

211. Describe how to style the hair for a client with uneven features.

____

____

212. Identify the rules to follow when styling the hair for clients with Negroid features.

____

213. Explain how to handle Negroid hair that has been straightened. ______________________________

214. Explain how to handle Negroid hair that has not been straightened.

____

215. Identify the rules to follow when styling the hair for clients with Oriental features.

____

216. Name the reason that Oriental hair requires more precise handling. ______________________________

STYLING FOR PEOPLE WHO WEAR GLASSES

217. List three factors that combine to accentuate the best features of clients who wear glasses.

1. ______________________ 2. ______________________

3. ______________________

218. List six good grooming rules for people who wear glasses.

1. ______________________
2. ______________________
3. ______________________
4. ______________________
5. ______________________
6. ______________________

219. Describe the type of glasses that should be worn by clients with round, oval, and square faces. ______________________

220. Identify the most flattering color for frames. ______________________

221. Describe how to style the hair for clients with round, oval, or square faces.

____

222. Describe the type of glasses that should be worn by clients with heart-shaped and diamond-shaped faces.

____

____

223. Describe the type of makeup that should be worn by clients with heart-shaped and diamond-shaped faces. ______________________

224. Describe how to style the hair for clients with heart-shaped or diamond-shaped faces.

____

225. Describe the type of glasses that should be worn by clients with small, narrow, or oval faces. ______________________

226. Describe the type of makeup that should be worn by clients with small, narrow, or oval faces. ______________________

227. Describe how to style the hair for clients with small, narrow, or oval faces.

____

228. Describe the type of glasses that should be worn by clients with pear-shaped faces.

____

229. Describe how to style the hair for clients with pear-shaped faces.

____

____

HAIR PARTINGS

230. List three characteristics of a good hair parting.

   1. ______________________________
   2. ______________________________
   3. ______________________________

231. Name which parting to use whenever possible. ______________________________

232. List three client characteristics that can be used to create a part.

   1. ______________ 2. ______________
   3. ______________

233. Name which parting to avoid when creating a lasting hairstyle.

   ____

234. List two partings most commonly used for children's bangs.

   1. ______________ 2. ______________

235. Explain the reason for using a triangular parting for children's bangs.

   ____

236. Describe why a diagonal part is used. ______________________________

237. Describe what a curved rectangular part is used for.

   ____

238. Describe what a center part is used for. ______________________________

239. Describe what a concealed part is used for. ______________________________

240. Describe what a side part is used for. ______________________________

241. Describe what facial types a center part is used for.

   ____

242. Explain the purpose of using partings on other parts of the head.

   ____

WORD REVIEW

back-combing
base
corn-row
natural part
profile
back-brushing
braid
indentation
pin curl
shaping
barrel curl
circle
mobility
plait
volume

MATCHING TEST

Insert the correct term or phrase in front of each definition.

back-combing	back-brushing	balance point
barrel curls	base	circle
clockwise	corn-rowing	counterclockwise
forward	interlocking movement	invisible braid
mobility	radial motion	reverse
ridge curls	semi-stand up curls	shaping
skip waves	stand up curls	stem
visible braid		

1. ______________________ the opposite direction of the movement of the hands of a clock
2. ______________________ the part of the pin curl that forms a complete circle
3. ______________________ another name for teasing or ratting
4. ______________________ the stationary, or immovable, foundation of a curl
5. ______________________ pin curls placed behind the ridge of a shaping or finger wave
6. ______________________ another name for a regular braid
7. ______________________ directed toward the face
8. ______________________ pin curls that provide for height in the finished hairstyle
9. ______________________ the same direction as the movement of the hands of a clock
10. ______________________ another name for ruffing
11. ______________________ a section of hair molded into a design to serve as a base for a curl or wave pattern
12. ______________________ a combination of finger waves and pin curl patterns
13. ______________________ another name for movement
14. ______________________ a type of French braiding incorporating narrow sections
15. ______________________ the section of the pin curl that gives the circle its direction, action, and mobility
16. ______________________ another name for the spot from which hair is directed to form a circular movement
17. ______________________ two directional rows of pin curls that produce clash
18. ______________________ another name for an inverted braid
19. ______________________ pin curls carved out of a shaping and pinned in a semi-standing position
20. ______________________ directed backward or away from the face

RAPID REVIEW TEST

Place the correct word in the space provided in each sentence below.

body	center	close
closed	coarse	curly
destroyed	equal	fine
firm	flatness	full
gaudy	hair	hot
large	length	no
open	overlap	parting
perpendicular	pie	pin curls
rollers	shaping	side
soft	small	splits
strong	tapered	tension
thick	three	two
volume	wave	wet
wide		

1. The shaping of a curl must not be ________________ when it is being combed or pinned.
2. It's important to examine the client's ___________ before starting the shampoo.
3. ___________ -stem curls produce tight, firm, long-lasting curls.
4. Becoming hairstyles are properly proportioned to individual ____________ structure.
5. _____________ -center pin curls produce even, smooth waves and uniform curls.
6. ____________ hair requires smaller rollers; _____________ hair needs larger rollers.
7. Stretching the hair strands and applying ______________ results in longer-lasting pin curls.
8. ______________ give more security to hair when it is ___________.
9. _____________ partings give an oval illusion to wide, round, and heart-shaped faces.
10. Clippies or hairpins may not be able to support the weight of coarse or ____________ hair.
11. Eyeglass frames should be up-to-date, with ____________ lenses.
12. Triangular-base pin curls are recommended along the front hairline to prevent _____________ in the finished hairstyle.
13. Cosmetologists must build volume on areas of the head where there is ______________.
14. The size of the curl determines the size of the ____________.
15. Cylinder-shaped rollers must be placed farther back from the point of distribution in ___________ -shaped sections.
16. Clips placed against the skin, ear, or scalp can become __________ during the drying process.
17. The hair is swept upward to give _____________ to the neck.

18. Back-combing provides a ____________ cushion; back-brushing provides a ____________ cushion.

19. When working against the natural crown ______________, hairstyles may not last.

20. Pin curls work best if the hair is properly ______________ and wound smoothly.

21. Rollers create a great deal of lift and ______________.

22. ______________-center curls produce waves that decrease in size toward the ends.

23. Eyeglass frames should not be exotic or ______________.

24. Clips are always inserted from the open end of the ______________.

25. Hair must be ____________ times the roller's diameter.

26. ____________-stem curls produce strong direction and weaker wave patterns.

27. ____________-set eyes are usually found on round and square faces; ____________-set eyes are usually found on long, narrow faces.

28. Oriental hair is usually ______________.

29. Uniform curl size occurs when hair sections are ______________.

30. Pin curls should ______________ one another.

MULTIPLE CHOICE TEST

Read each statement carefully, then write the letter representing the word or phrase that correctly completes the statement on the blank line to the right.

1. Before shampooing, cutting, or styling hair, it is necessary to remove
   - a) color
   - b) curl
   - c) tangles
   - d) texture ________

2. To achieve full volume, the roller sits
   - a) on its base
   - b) half off its base
   - c) off its base
   - d) overdirected off its base ________

3. A center part is most suitable for clients whose noses are
   - a) turned up
   - b) prominent
   - c) crooked
   - d) wide and flat ________

4. Braiding must be done with
   - a) no tension
   - b) loose tension
   - c) even tension
   - d) inconsistent tension ________

5. The face shape considered to be ideal is the
   - a) oblong
   - b) oval
   - c) round
   - d) square ________

6. Hair that has been straightened is set on
   - a) large rollers
   - b) medium rollers
   - c) small rollers
   - d) brush rollers ________

7. A long, thin neck is minimized by keeping hair long and full at the
   a) forehead   b) temples
   c) cheekbones   d) nape   ________

8. Pin curls carved out of a shaping are referred to as
   a) shaped curls   b) carved curls
   c) horizontal curls   d) vertical curls   ________

9. Corn-rowing works well with hair that is
   a) straight   b) slightly wavy
   c) overly straightened   d) overly curly   ________

10. The cosmetologist's objective is to accentuate a client's
   a) best features   b) worst features
   c) blemishes   d) defects   ________

11. Eyeglass frames should match the color of the
   a) eyes   b) skin
   c) hair   d) clothes   ________

12. After removing rollers and clips, the hair is brushed thoroughly to
   a) pull out the hair   b) mess up the hair
   c) relax the set   d) straighten the set   ________

13. Facial shape is determined by the position and prominence of facial
   a) bones   b) muscles
   c) nerves   d) blood vessels   ________

14. The angle at which hair is held from the head determines the
   a) pivot   b) stem
   c) circle   d) base   ________

15. The cosmetologist's objective for round and square facial types is to create the illusion of
   a) length to the face   b) width to the temples
   c) width to the forehead   d) width to the jawline   ________

16. Braids are neater and longer-lasting when done on hair that is
   a) dripping wet   b) damp
   c) dry   d) oily   ________

17. The head shape considered to be ideal is the
   a) oblong   b) oval
   c) round   d) square   ________

18. Finding the natural part requires combing the hair straight back, placing the palm on the head, and pushing
   a) backward   b) forward
   c) toward the left side   d) toward the right side   ________

19. Asymmetrical styles are most suitable for clients whose noses are
   a) turned up   b) prominent
   c) crooked   d) wide and flat   ________

20. Pin curls must be placed in the direction in which they will be

a) combed
b) sprayed
c) tinted
d) back-brushed
____

21. The cosmetologist's objective for the diamond-shaped face is to reduce the width across the

a) forehead
b) cheekbones
c) temples
d) jawline
____

22. The face is divided into

a) two zones
b) three zones
c) four zones
d) five zones
____

23. Triangular partings distribute more hair to the

a) nape area
b) crown area
c) top area
d) temple area
____

24. Invisible braiding is done by overlapping the strands

a) under
b) on top
c) on the left side
d) on the right side
____

25. The profile considered to be ideal is

a) straight
b) round
c) concave
d) convex
____

26. To achieve a small amount of lift, the roller sits

a) on its base
b) half off its base
c) off its base
d) overdirected off its base
____

27. Back-combing provides additional

a) curl
b) volume
c) color
d) length
____

28. The cosmetologist's objective for the oblong facial type is to make the face appear

a) longer and wider
b) longer and narrower
c) shorter and wider
d) shorter and narrower
____

29. Hair partings must be neat and

a) overdirected
b) underdirected
c) crooked
d) straight
____

30. When working with rollers, cosmetologists must wrap hair ends

a) carelessly
b) sloppily
c) smoothly
d) when dry
____

Date ________________

Rating ________________

Text Pages 163–186

THERMAL HAIRSTYLING

INTRODUCTION

1. Explain why thermal waving is still known as marcel waving.

____

2. Describe thermal waving and curling.

____

____

THERMAL IRONS

3. Explain how the heat of thermal irons is controlled.

____

4. Explain why irons must be made of the best-quality steel.

____

5. Name and describe the two parts of the styling portion of the irons.

   1. ____

   2. ____

6. List three classifications of thermal irons.

   1. ____________________ 2. ____________________

   3. ____________________

7. Explain why electric vaporizing irons should not be used on pressed hair.

____

8. List two factors that govern temperature settings for irons.

   1. ____

   2. ____

9. Name two types of hair that should be curled and waved with lukewarm irons.

   1. ____________________ 2. ____________________

10. Name two types of hair that can tolerate more heat than fine hair.

   1. ____________________ 2. ____________________

11. Explain why thermal irons should not be used on chemically treated hair.

____

12. Specify what is used to test the heated irons. _______________

13. Describe how to test the irons.

____

14. Explain what happens if the irons are too hot. _______________

15. List three types of hair that withstand less heat than normal hair.

1. _______________ 2. _______________

3. _______________

16. Describe how to remove dirt or grease from thermal irons.

____

17. Explain how to remove rust and carbon from irons.

____

18. Describe why the joint of the irons should be oiled.

____

19. Explain what happens when irons are overheated.

____

20. Describe how to hold the irons.

____

____

____

21. Describe the comb that should be used with thermal irons.

____

____

22. Describe how to hold the comb.

____

____

____

23. Describe the best way to practice using thermal irons.

____

24. Specify what is used to roll the handles in either direction.

____

THERMAL WAVING WITH CONVENTIONAL THERMAL (MARCEL) IRONS

25. Explain how to determine whether the first wave will be a left-going or a right-going wave. ____

26. List the eight steps required for a left-going wave.

 1. ____
 2. ____
 3. ____
 4. ____
 5. ____
 6. ____
 7. ____
 8. ____

27. Describe how to direct hair for a right-going wave.

____

28. Explain why a small section of the waved strand is included when picking up unwaved hair in the comb. ____

29. Describe what happens when movements of comb and irons are not the same for waving the second strand as for waving the first. ____

THERMAL CURLING WITH ELECTRIC THERMAL IRONS

30. Identify what is eliminated when styling hair with thermal irons.

____

31. List four advantages of thermal waving straight hair.

1. ____________________ 2. ____________________
3. ____________________ 4. ____________________

32. Explain the advantage of thermal waving pressed hair.

____

33. Explain the advantage of thermal waving wigs and hairpieces.

____

34. List two different ways to open the clamp of the irons.

1. ____________________ 2. ____________________

35. Explain how to develop a smooth rotating movement with the irons.

____

36. Describe how to release the hair from the irons.

____

37. Explain how to insure that the end of the strand will be in the center of the curl.

____

38. Describe how to remove the curl from the irons.

____

39. Name the implement used to protect the client's scalp from burns. ____________________

THERMAL IRONS CURLING METHODS

40. List the four steps required to prepare hair for curling with thermal irons.

1. ____
2. ____
3. ____
4. ____

41. Describe the size of the base usually required for curling short hair.

____

42. Explain why hair must be combed smooth.

____

43. Name the part of the irons that must be on top. ______________________________

44. Specify the distance from the scalp at which to insert irons when curling short hair.

____

45. Explain why irons are held on hair for a few seconds after inserting. ______________

46. Describe how to hold the ends of the hair strand.

____

47. Describe the amount of tension required to hold the hair ends.

____

48. Explain how to prevent binding of the hair in the irons.

____

49. Specify the distance from the scalp at which to insert irons when curling medium-length hair. ______________________________

50. Explain how long to hold irons on the hair to heat it. ______________________

51. Specify the distance at which to slide irons from the scalp. __________________

52. Name the part of the irons that must be on top. ______________________________

53. Explain how to handle the strand after turning irons downward.

____

____

54. Describe how to enlarge the curl. ______________________________

55. Explain where to insert the ends of the curl.

____

56. Describe how to move the ends of the strand into the center of the curl.

____

57. Explain how to even out the distribution of the hair in the curl.

____

58. Describe how to protect the client's scalp during the curling process.

____

59. List two reasons for a final revolution of irons inside the curl.

1. ______________________ 2. ______________________

60. Describe the action of the irons and comb to remove hair from irons.

____

____

61. Specify the distance from the scalp at which to insert irons when curling long hair.

____

62. Specify the direction in which to pull the hair over the rod.

___

63. Explain how long to hold irons on the hair to heat it. ___

64. Identify the amount of tension needed to hold the hair strand.

___

65. Name the direction in which to roll the irons. ___

66. Name the part of the irons toward which to draw the hair strand.

___

67. Explain the purpose of pushing the irons forward and pushing the hair with the left hand.

___

___

68. List two reasons for rotating irons several times after completing curl.

1. ___

2. ___

69. Describe spiral curls. ___

70. Describe end curls. ___

VOLUME THERMAL IRON CURLS

71. Explain what determines the type of volume curls to use. ___

72. Describe the volume-base curl.

___

73. Explain how to hold the hair strand to form a volume-base curl. ___

74. Describe the full-base curl. ___

75. Explain how to hold the hair strand to form a full-base curl. ___

76. Describe the half-base curl. ___

77. Explain how to hold the hair strand to form a half-base curl. ___

78. Describe the off-base curl. ___

79. Explain how to hold the hair strand to form an off-base curl. ___

80. Draw a line from each curl base to the correct description.

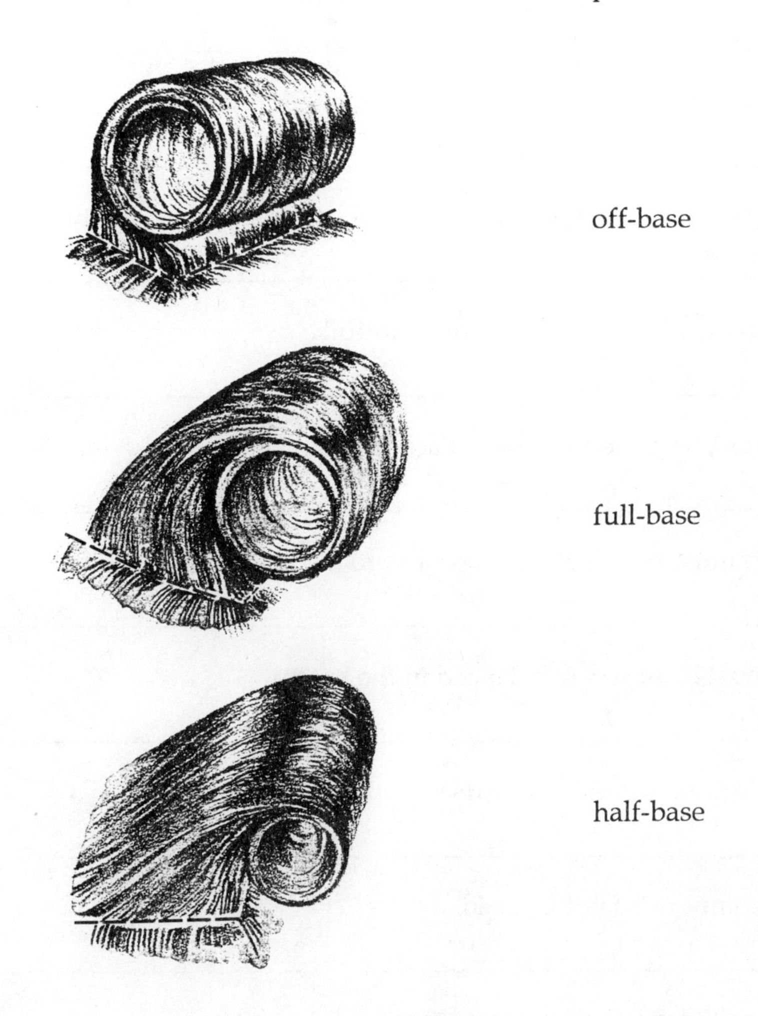

FINISHED THERMAL CURL SETTINGS

81. Explain how to obtain best results when thermal setting.

____

STYLING THE HAIR AFTER A THERMAL CURL OR WAVE

82. Identify whose wishes to follow when styling a client's hair. ____________________

83. Explain how to style a client's hair after thermal waving or curling.

____

____

SAFETY MEASURES

84. Explain why irons should not be overheated.

____

85. Explain the necessity of testing the temperature of the irons on tissue paper before placing them on the hair. ______

86. Explain why the fumes of the irons must not be inhaled.

______

87. Explain the necessity of not placing irons near the face to test for temperature.

______

88. Explain why cosmetologists must handle irons carefully.

______

89. Explain why irons must be placed in a safe place to cool.

______

90. Explain why handles must not be placed too close to the heater.

______

91. Explain why irons must be properly balanced in the heater.

______

92. Explain why celluloid combs must not be used in thermal curling.

______

93. Explain why metal combs must not be used.

______

94. Explain why combs with broken teeth must not be used.

______

95. Explain why a comb is placed between the scalp and thermal irons.

______

96. Explain why the client's hair must be clean for thermal curling.

______

97. What must be done to thick or bulky hair before thermal curling.

______

98. Explain why hair ends must not protrude over the irons. ______

99. Explain why vaporizing irons should not be used on pressed hair.

______

100. Explain why thermal irons should not be used on chemically straightened hair.

______

BLOW-DRY STYLING

101. Describe blow-dry styling.

____

102. List two advantages of blow-dry styling.

1. ____

____

2. ____

103. List two basic techniques used in blow-dry styling.

1. ____________________ 2. ____________________

104. Name the type of hair that must be handled with care when blow-dry styling.

____

105. Describe how to handle hair that has loss of elasticity.

____

EQUIPMENT, IMPLEMENTS, AND MATERIALS

106. List the equipment, implements, and materials required for blow-dry styling.

1. ____________________ 2. ____________________
3. ____________________ 4. ____________________
5. ____________________ 6. ____________________

THE BLOW DRYER

107. Describe the blow dryer (without attachments).

____

108. List the five main parts of a blow dryer.

1. ____________________ 2. ____________________
3. ____________________ 4. ____________________
5. ____________________

109. Explain what a blow dryer does when in operation.

____

COMBS AND BRUSHES

110. Name two types of combs used for blow waving and air waving.

____

111. Explain why metal combs are preferred by some stylists.

____

____

112. Describe the type of brush used in blow-dry styling. ______________________________

113. Explain the results of using a brush with a smaller diameter.

____

COSMETICS USED IN BLOW-DRY STYLING

114. Explain the reason for applying styling lotions to hair after shampooing.

____

115. Describe how to apply styling lotions.

____

116. Explain the advantage of the coating substance in styling lotions.

____

117. Explain how often hair conditioners are used as a corrective treatment for dry and brittle hair. ______________________________

118. Name the effects of excessive blow-drying.

____

119. Identify the type of conditioner to use for blow-dry styling. ______________________________

120. Describe the purpose of hair sprays. ______________________________

BLOW CURLING WITH ROUND BRUSH

121. Describe the type of hair most suitable for blow curling.

____

122. Explain how hair should be shaped prior to blow curling.

____

123. Identify the type of shaping that makes blow curling difficult.

____

124. List the eight steps required to blow curl a client's hair.

1. ______________________________
2. ______________________________
3. ______________________________
4. ______________________________
5. ______________________________
6. ______________________________

   ____
7. ______________________________
8. ______________________________

REMINDERS AND HINTS ON BLOW-DRY STYLING

125. Explain how to achieve the best results from blow-dry styling.

____

126. Explain why partial towel-drying chemically treated hair prior to blow-drying is recommended. _______________

127. Explain how to spread the styling lotion evenly. _______________

BLOW-DRYING THE HAIR

128. Explain for what purpose hot air is directed straight onto the head.

____

129. Explain how to direct the air for blow-dry styling.

____

130. List three reasons for directing the air flow to the top half of the brush in a back-and-forth movement.

1. _______________ 2. _______________
3. _______________

131. Describe how to direct the blower.

____

132. Explain how to avoid severe scalp burns.

____

133. Describe how to cool the hair before combing it out.

____

134. Explain why the scalp must be thoroughly dry when the hairstyle is completed.

____

135. List two benefits of spraying the completed style with hair spray.

1. _______________ 2. _______________

BRUSHES AND COMBS

136. Explain what implement actually does the styling. _______________

137. Describe the purpose of the blow dryer. _______________

138. List two factors that determine the size of the styling brush.

1. _______________ 2. _______________

139. Compare the sizes of brushes used to style short, medium, or longer hair.

____

____

140. Explain how to avoid scalp burns. ______________________________

BLOW DRYER

141. Explain what happens when the blow dryer is not kept free of dirt or hair.

____

142. Explain why the air intake at the back of the dryer must be kept clear at all times.

____

____

BLOW-DRYING TECHNIQUES

143. Describe how to give the crown hair a slight lift.

____

144. Explain how to achieve a page boy effect. ______________________________

145. Describe how to create a smooth top with flip.

____

____

AIR WAVING

146. Name a technique similar to air waving. ______________________________

147. List two implements used to air wave hair.

1. ______________ 2. ______________

148. List three services that precede air waving.

1. ______________ 2. ______________
3. ______________

149. Explain why hair is combed in the direction of the wave desired.

____

150. Explain how long to comb the hair with the air waver.

____

151. Explain why hair must be slightly damp.

____

SHAPING HAIR WITH COMB

152. Name the area of the head where air waving is started. ______________________

153. Specify the distance from the part at which the comb is inserted. ______________________

154. Explain how to form the first ridge.

____

____

____

155. Name the area of the head where the second ridge is started.

____

156. Name another implement that can be used in conjunction with the air waver comb.

____

SAFETY PRECAUTIONS

157. List two safety precautions associated with the styling dryer.
    1. ______________________
    2. ______________________

158. Explain how to avoid scalp burns when working with a metal comb.

____

159. Explain why the scalp must be thoroughly dry when curling or waving is completed.

____

WORD REVIEW

celluloid
fishhooks
lubricant
rod
temper
volume
croquignole
flammable
marcel
shell
thermal
end curls
loop
nonflammable
spiral
vaporizing

MATCHING TEST

Place the correct term or phrase in front of each definition.

air waving
end curl
inner edge
shell
volume-base curl
blow-dry styling
full-base curl
off-base curl
spiral curl
croquignole curl
half-base curl
rod
thermal waving

1. ______________________ a curl providing maximum lift or volume

2. ______________________ a hanging curl suitable for long hairstyles

3. ______________________ a curl providing full volume

4. ______________________ the technique of drying and styling hair in one operation

5. ______________________ the part of the thermal irons that is a perfectly round, solid steel bar

6. ______________________ a curl providing only slight lift or volume

7. ______________________ the art of waving hair using conventional marcel irons

8. ______________________ a curl used to give a finished appearance to hair ends

9. ______________________ a curl providing moderate lift or volume

10. ______________________ the part of the thermal irons that is perfectly round with inside grooved

RAPID REVIEW TEST

Place the correct word in the space provided in each sentence below.

back-and-forth	balanced	broken
brush	burns	celluloid
clear	coating	cooled
comb	damaged	elasticity
ends	face	fine
flexible	fumes	growth
guide	loosens	lubricant
metal	normal	oil
opposite	porous	rough
rubber	safe	same
scalp	smoothes	temper
tight	towel-dried	wet

1. Thermal curling eliminates styling hair while it is __________.
2. A ____________ is used between the scalp and thermal irons to protect the client.
3. Placing the handles of the irons too close to the heater may result in ____________ when removing the irons.
4. The hot air of styling dryers must be directed in a ____________________ movement.
5. When picking up unwaved hair in the comb, including a small section of the waved strand serves as a ____________ to the formation of the new wave.
6. Fine, lightened, or badly damaged hair can withstand less heat than _____________ hair.
7. _____________ combs should not be used for thermal curling because they become hot and burn the scalp.
8. The air intake at the back of the dryer must be kept ____________ at all times.
9. Whether a wave will be right-going or left-going is determined by the hair's natural _____________.
10. Inhaling _____________ from the irons is injurious to the lungs.
11. Before blow-drying chemically treated or damaged hair, it must be _________________.
12. Overheated irons usually lose their _____________ .

13. Styling lotions contain a ______________ substance that gives hair more body.

14. Combs with ______________ teeth should not be used for thermal curling because they can break or split the hair or injure the scalp.

15. Blow-dry styling involves directing the air from the scalp area to the hair ____________.

16. Hot irons should not be placed near the ___________ to test for temperature.

17. Chemically treated or damaged hair has loss of ________________.

18. The joint of thermal irons needs __________ for greater facility in movement.

19. Hot irons should be placed in a ___________ place to cool.

20. A final revolution of the thermal irons inside the curl ________________ the ends and ______________ the hair away from the irons.

21. In blow-drying, brushes with the smallest diameter produce ____________ curls.

22. Hair conditioners containing a _______________ are recommended for hair damaged by excessive blow drying.

23. Hair should be thoroughly _____________ before it is combed out.

24. Irons that are improperly _______________ in the heater may fall and injure someone.

25. Hot air is directed straight onto the head only for _____________ blow drying.

26. Usually, coarse and gray hair can tolerate more heat than ___________ hair.

27. The style and length of hair determine the size of the styling ____________.

28. _______________ combs should not be used for thermal curling because they are flammable.

29. In blow-drying, the hot air is directed away from the client's ____________ to avoid severe burns.

30. A curl is removed from thermal irons by drawing the irons in one direction while drawing the comb in the _______________ direction.

MULTIPLE CHOICE TEST

Read each statement carefully, then write the letter representing the word or phrase that correctly completes the statement on the blank line to the right.

1. The rolling movement of the irons is accomplished by using only the

   a) thumb b) fingers
   c) wrist d) arm ________

2. Thermal curling is not recommended for

   a) straight hair b) pressed hair
   c) wigs and hairpieces d) chemically treated hair ________

3. The creation of hairstyles without time-consuming setting, drying, and combing out is known as

   a) blow-dry styling b) roller setting
   c) wet setting d) wet hairstyling ________

4. Combs used with thermal irons should have

   a) missing teeth b) broken teeth
   c) fine teeth d) coarse teeth ________

5. A hairstyle will not hold if the scalp is

   a) dry b) damp
   c) tight d) flexible ________

6. Thermal waving and curling is the art of waving and curling hair that is

   a) straight or pressed b) wavy
   c) curly d) overly curly ________

7. Dirt and grease are removed from thermal irons by washing in a soap solution containing a few drops of

   a) detergent b) disinfectant
   c) alcohol d) ammonia ________

8. Air waving is a technique similar to

   a) finger waving b) permanent waving
   c) thermal waving d) blow waving ________

9. The temperature of heated thermal irons is tested on a

   a) dark towel b) damp cloth
   c) strand of hair d) piece of tissue paper ________

10. Styling lotions are applied to hair after shampooing to make hair more manageable for

    a) tinting b) lightening
    c) blow curling or waving d) chemical relaxing ________

11. Combs used for thermal curling should be made of

    a) hard rubber b) soft rubber
    c) metal d) plastic ________

12. Lukewarm irons should be used on hair that is

a) gray
b) resistant
c) coarse
d) tinted or lightened ________

13. The technique of curling hair with thermal irons was developed by

a) Charles Nessler
b) Marcel Grateau
c) Rousseau Sabouraud
d) Pierre Marcel ________

14. Styling hair with an air waver requires that hair be

a) oily
b) dry
c) lightly damp
d) dripping wet ________

15. The croquignole curl requires the formation of

a) one loop
b) two loops
c) three loops
d) four loops ________

16. To insure a good thermal curl or wave, the client's hair must be

a) dirty
b) greasy
c) clean
d) damp ________

17. To receive a blow-curling service, the hair should be shaped with

a) blunt-cut ends
b) tapered ends
c) frizzy ends
d) fishhook ends ________

18. Using electric vaporizing thermal irons on pressed hair can cause hair to revert to its natural

a) overly curly state
b) elasticity
c) color
d) porosity ________

19. Thermal irons are inserted onto hair about

a) 1" from scalp
b) 1 1/2" from scalp
c) 2" from scalp
d) 2 1/2" from scalp ________

20. Each client's hair should be styled according to the wishes of the

a) cosmetologist
b) salon manager
c) client
d) client's spouse ________

21. Clamps on thermal irons are opened with the little finger only or little finger and

a) thumb
b) index finger
c) middle finger
d) ring finger ________

22. Combs used for blow waving and air waving should be made of hard rubber or

a) plastic
b) metal
c) celluloid
d) soft rubber ________

23. Thermal irons provide an even heat that is completely controlled by the

a) manufacturer
b) client
c) salon manager
d) cosmetologist ________

24. Using thermal irons on chemically treated hair can cause hair to

a) break off
b) become healthier
c) darken
d) lighten ________

25. Before thermal curling thick and bulky hair, it should be

a) lightened
b) darkened
c) thinned and tapered
d) permanent waved

____

Date ______________

Rating ______________

Text Pages 187–224

PERMANENT WAVING

CHOOSING THE RIGHT PERMING TECHNIQUE

1. Name two ways to obtain information necessary to select the perm product and technique appropriate for each client.
   1. ______________
   2. ______________
2. Identify a characteristic necessary for successful perming. ______________
3. List three pre-perming skills that require considerable practice.
   1. ______________ 2. ______________
   3. ______________
4. Explain how to determine the client's expectations.

   ____
5. List four points to cover during the consultation.
   1. ______________
   2. ______________
   3. ______________
   4. ______________
6. Identify the amount of time it takes to consult with a perm client. ______________
7. Name three reasons why the consultation is worth the time.
   1. ______________

      ____
   2. ______________

      ____
   3. ______________

      ____
8. Explain what to do with the information obtained during the client consultation.

   ____

PRE-PERM ANALYSIS

9. List three factors determined by the pre-perm analysis.

   1. ______________________________
   2. ______________________________
   3. ______________________________

10. Explain what to do when abrasions, irritations, or open sores are present on the scalp.

    ____

11. List five physical characteristics of the hair that need analyzing.

    1. ______________ 2. ______________
    3. ______________ 4. ______________
    5. ______________

12. List four classifications of chemicals that, used previously on the hair, affect the choice of the perm.

    1. ______________________________
    2. ______________________________
    3. ______________________________
    4. ______________________________

13. Explain the result of an incorrect pre-perm analysis.

    ____

14. Describe porosity. ______________________________

15. List two factors that have a direct relationship to the hair's porosity.

    1. ______________________________
    2. ______________________________

16. Name the one factor that processing time is most dependent on. ______________

17. List six factors that affect hair porosity.

    1. ______________ 2. ______________
    3. ______________ 4. ______________
    5. ______________ 6. ______________

18. List three general classifications of hair that absorb liquids readily.

    1. ______________________________
    2. ______________________________
    3. ______________________________

19. Describe how to check for porosity on a single strand of hair.

    ____

    ____

20. Explain how to determine if hair is not porous.

____

21. Explain how to determine if hair is porous.

____

22. Describe hair with poor porosity. _______________

23. Identify the amount of processing time and the strength of waving lotion required for hair with poor porosity. _______________

24. Describe hair with good porosity.

____

25. Identify the amount of processing time required for hair with good porosity.

____

26. Describe porous hair.

____

____

27. Identify the amount of processing time and type of perm required for porous hair.

____

28. Describe over-porous hair. _______________

29. Explain how to treat over-porous hair.

____

30. Describe hair that is unevenly porous.

____

31. List two reasons why a pre-wrap lotion is recommended for unevenly porous hair.

    1. _______________

    2. _______________

32. Define texture. _______________

33. Differentiate between fine and coarse hair.

____

34. Compare the importance of texture when estimating processing times for fine and coarse hair of equal and unequal porosity.

____

____

____

35. Define elasticity. _______________

36. Describe how to test for elasticity. _______________

37. Explain how to determine if hair has little or no elasticity.

____

38. List two other signs of poor elasticity.

1. ______________ 2. ______________

39. Explain why hair completely lacking in elasticity will not take a satisfactory permanent wave. ______________________________

40. Describe hair with good elastic qualities.

____

41. Define density, or thickness. ______________________________

42. Explain what is determined by hair's density. ______________________________

43. Compare the partings required for thick and for thin hair.

____

____

44. Explain what happens when too much hair is wrapped on the rods.

____

45. Explain what happens when hair is stretched or pulled toward the rod.

____

46. Identify the hair length ideal for perming. ______________________________

47. Identify the number of turns hair should wrap around the rod. ______________

48. Explain why smaller partings must be used to perm hair longer than 6."

____

PERM SELECTION

49. Explain what the choice of perm depends on.

____

____

50. List two hair types that require an alkaline lotion wrap or alkaline water wrap perm.

1. ______________ 2. ______________

51. List three hair types that require an alkaline water wrap or acid-balanced perm.

1. ______________ 2. ______________

3. ______________

52. List five hair types that require an acid-balanced perm.

1. ______________ 2. ______________

3. ______________ 4. ______________

5. ______________

PRE-PERM SHAMPOOING

53. Explain why pre-perm shampoos are recommended for optimal perm results.

____

54. List four causes for a coating on the hair.
    1. _______________
    2. _______________
    3. _______________
    4. _______________

55. Explain why hair must be free of all coatings before starting the perm.

____

____

56. Explain why vigorous brushing, combing, pulling, or rubbing must be avoided prior to a perm. _______________

57. Identify the length of time to leave shampoo on the hair if it is extremely coated.

____

58. Describe the purpose of a thorough rinsing.

____

PRE-PERM CUTTING OR SHAPING

59. Explain why hair is texturized or thinned after the perm.

____

60. Explain how to shape hair when client wants a completely new style.

____

PERM RODS

61. Name what the size of the rod controls. _______________

62. Name the material that most perm rods are made of. _______________

63. Identify the range in the diameter of perm rods. _______________

64. List three colors that designate small perm rods.
    1. _______________ 2. _______________
    3. _______________

65. List three colors that designate medium-size perm rods.
    1. _______________ 2. _______________
    3. _______________

66. List four colors that designate large perm rods.

1. ______________________ 2. ______________________

3. ______________________ 4. ______________________

67. List three lengths of perm rods.

1. ______________________ 2. ______________________

3. ______________________

68. Describe concave rods.

____

____

69. Describe the type of curl produced by the concave rod.

____

70. Describe straight rods. ______________________________

71. Describe the type of curl produced by the straight rod.

____

72. Explain how the hair is secured on the rods.

____

73. List two factors to consider when selecting rod size.

1. ______________________ 2. ______________________

74. List three factors that determine success in creating the style.

1. ______________________ 2. ______________________

3. ______________________

75. List three hair characteristics important to selection of rod sizes.

1. ______________________ 2. ______________________

3. ______________________

76. Name the determining factor in the choice of rods. ______________________

77. Describe the partings and rods required for hair of coarse texture and good elasticity.

____

78. Describe the partings and rods required for hair of medium texture and average elasticity.

____

79. Describe the partings and rods required for hair of fine texture and poor elasticity.

____

80. Describe the partings and rods required for hair in nape area.

____

81. Describe the partings and wrap required for long hair.

____

SECTIONING AND PARTING

82. Define sectioning.

____

____

83. Explain why sectioning makes the work easier.

____

84. Define parting, or blocking. ___________________________

85. Explain why blocking is helpful.

____

____

86. List the five guidelines that help insure uniform blockings.
    1. ____
    2. ____
    3. ____
    4. ____
    5. ____

       ____

WRAPPING PATTERNS

87. Identify what is determined by the wrapping pattern.

____

88. Explain when to use the single halo wrap. _______________________

89. Explain when to use the double halo wrap. _______________________

90. Explain when to use the straight back wrap.

____

91. Describe how to create bangs on the forehead.

____

92. Explain when to use the dropped crown wrap.

____

93. Explain when to use the spiral wrap. ___________________________

94. Explain when to use the stack perm. ___________________________

WRAPPING THE HAIR

95. Describe how hair must be wrapped to create a uniform wave or curl pattern.

____

96. Explain what happens when hair is tightly wrapped.

____

____

97. Describe base.

____

98. Explain how to hold the hair strand for an on-base curl. _______________

99. Describe the types of hairstyles requiring on-base curls.

____

100. Explain how to hold the hair strand for an off-base curl. _______________

101. Describe the types of hairstyles requiring off-base curls.

____

102. Explain how to hold the hair strand for a one-half off-base curl.

____

103. Describe the types of hairstyles requiring one-half off-base curls.

____

104. Describe end wraps or end papers.

____

105. List three reasons for using end wraps.

   1. _______________
   2. _______________
   3. _______________

106. List three methods of end wrap application.

   1. _______________ 2. _______________
   3. _______________

107. Describe the desired amount of moisture on the hair for wrapping.

____

108. Explain what happens when hair partings are too long.

____

109. Explain what to do if hair becomes dry while wrapping. _______________

110. List the five steps required for using the double end-paper wrap.

1. ______________________________

2. ______________________________

3. ______________________________

4. ______________________________

5. ______________________________

111. Number the illustrations (from 1–5) in the correct order for winding a hair strand.

______

______

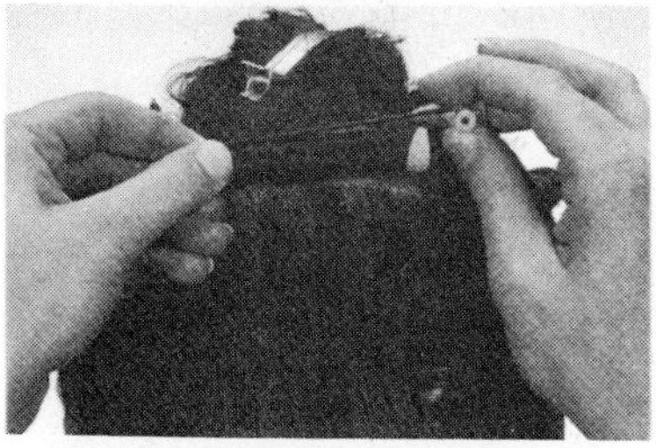

______

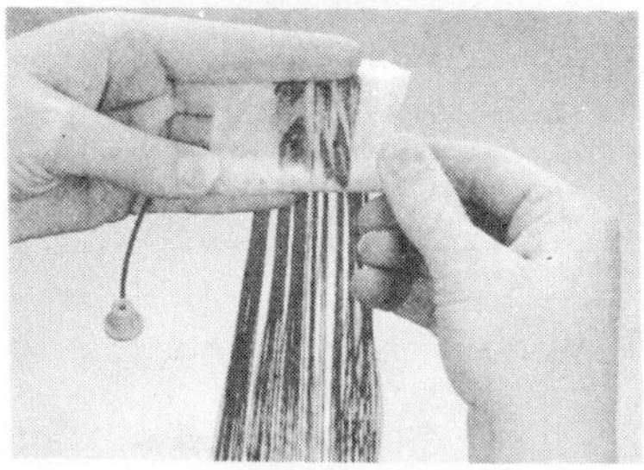

______

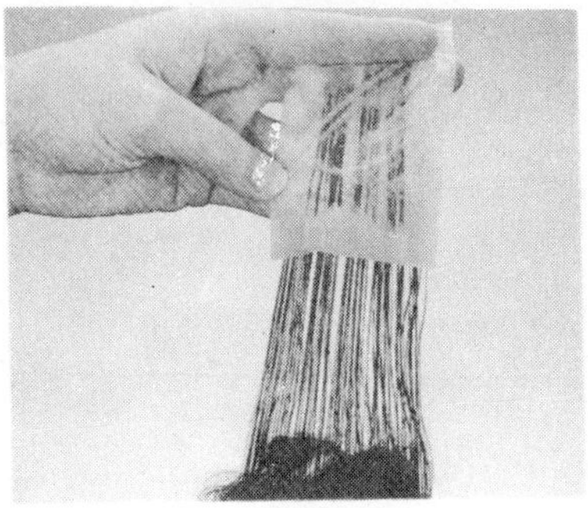

______

112. Explain how to prevent the band from causing hair breakage.

____

113. Describe how the single end-paper wrap differs from the double end-paper wrap.

____

____

114. Describe how the book end wrap differs from the double end-paper wrap.

____

____

115. Explain why the piggyback method of wrapping is especially suitable for extra long hair.

____

____

116. Compare the results of wrapping the piggyback with larger rods to wrapping with small or medium rods.

____

____

117. List the ten steps required for wrapping in the piggyback method.

1. _______________
2. _______________
3. _______________
4. _______________
   ____
5. _______________
6. _______________
7. _______________
   ____
8. _______________
9. _______________
10. _______________

118. List two reasons to avoid bulkiness of hair on the rods.

1. _______________
   ____
2. _______________

119. Explain how to avoid fishhook ends.

____

____

PRELIMINARY TEST CURLS

120. Explain the purpose of preliminary test curls (or pre-perm test curls).

____

121. List five hair conditions for which preliminary test curls are advisable.

1. ______________ 2. ______________
3. ______________ 4. ______________
5. ______________

122. List two pieces of additional information gained from preliminary testing.

1. ______________________________
2. ______________________________

123. List the six steps required for preliminary test curling.

1. ______________________________
2. ______________________________

____

3. ______________________________
4. ______________________________

____

5. ______________________________
6. ______________________________

124. Specify the number of turns to unwind a curl when checking a test curl.

____

125. Describe how to hold the hair when checking a test curl.

____

126. Describe how to move the rod when checking a test curl.

____

127. Describe the final test curl pattern.

____

128. Explain why the "S" formation on fine, thin hair is more difficult to read than the "S" formation on coarse, thick hair.

____

____

____

129. Explain why long hair may produce a wider scalp wave than short hair.

____

130. List four steps necessary to complete the test curls.

1. ____
2. ____
3. ____
4. ____

131. Explain what to do if test curls are overprocessed.

____

132. Explain what to do if the test curl results are good.

____

133. Describe overprocessed hair. ____

134. Explain why overprocessed hair cannot be combed into a suitable wave pattern.

____

135. Describe what should be done for overprocessed hair.

____

136. List three causes of overprocessing.

1. ____
2. ____
3. ____

137. Explain the cause of underprocessing.

____

138. Describe underprocessed hair.

____

____

139. Draw a line from each illustrated hair strand to the correct description.

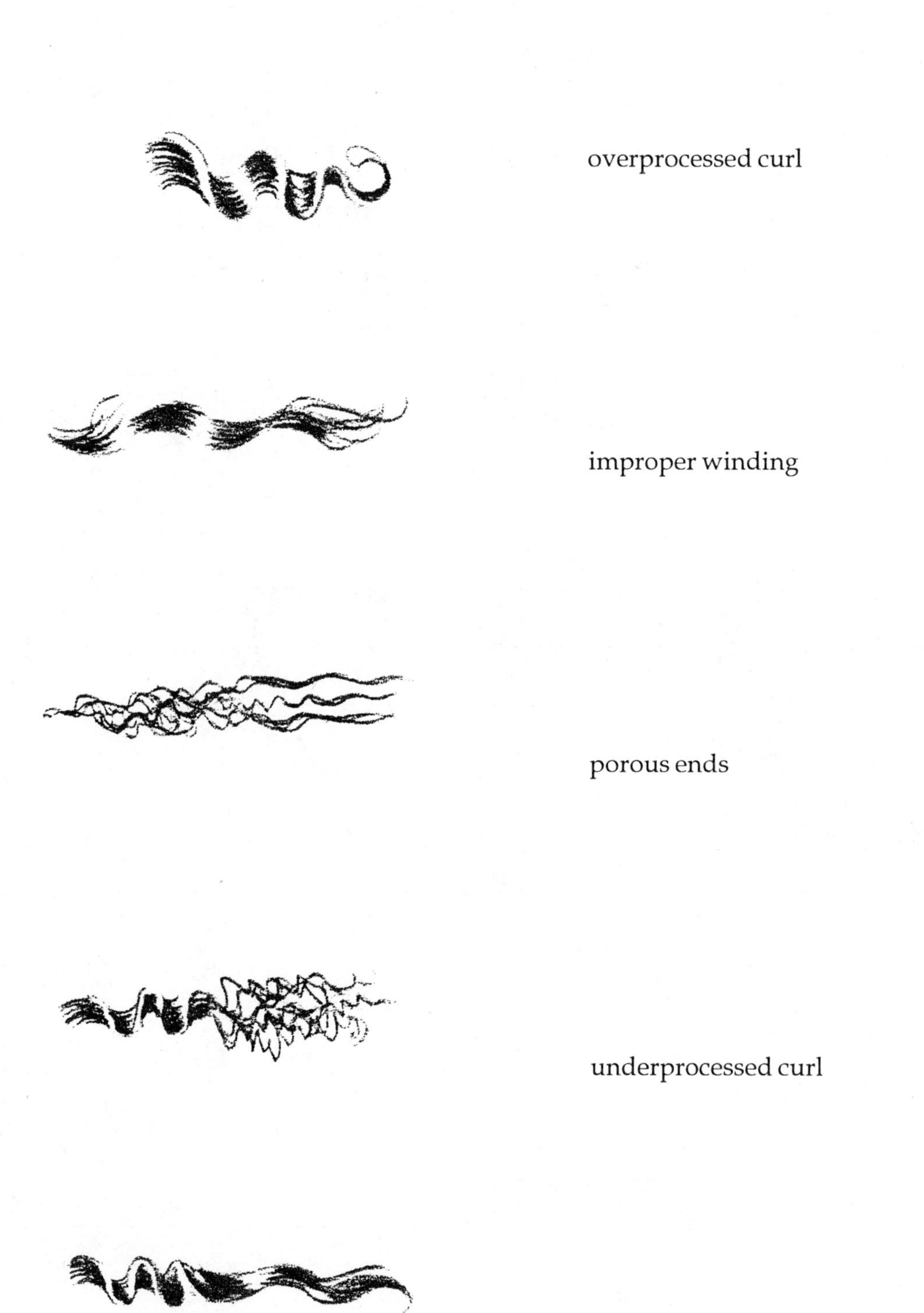

140. List three conditions necessary to re-perm underprocessed hair.

1. ________________ 2. ________________
3. ________________

141. Explain why lotions used for perming must be used carefully.

____

142. List seven precautions that must be followed when perming.

1. ________________________________
____
2. ________________________________
3. ________________________________
____
4. ________________________________
5. ________________________________
____
6. ________________________________
7. ________________________________

PERMING TECHNIQUES

143. List eight supplies required for permanent waving.

1. ________________ 2. ________________
3. ________________ 4. ________________
5. ________________ 6. ________________
7. ________________ 8. ________________

144. Explain how often cosmetologists should read the directions that accompany every perm.

____

APPLYING WAVING LOTION

145. Identify a safety precaution that prevents waving lotion from coming into contact with the skin. ________________________________

146. Explain how to achieve a minimum of dripping.

____

147. Describe what to do with cotton after application of waving lotion.

____
____

148. Describe how to apply waving lotion to wound curls.

____

____

149. Define processing time.

____

____

150. Name two factors that determine processing time.

1. _______________ 2. _______________

151. Explain the importance of accurately timing the perm process.

____

152. Explain why a record of previous processing times should be used only as a guide.

____

153. List five instances when a second resaturation of all the rods may be necessary.

1. _______________
2. _______________
3. _______________
4. _______________
5. _______________

154. Explain why wave development must be watched closely after a reapplication of lotion.

____

155. Describe how to correctly place a plastic cap over the rods.

____

156. Explain what happens if the plastic cap is too loose or does not cover all the rods.

____

157. Identify the heat setting and amount of airflow required for a preheated dryer.

____

158. Specify the length of time required for the dryer to warm up before placing the client under it. _______________

159. Explain what happens if the client sits in a draft or too close to an air conditioner.

____

160. Explain how often optimum curl development occurs.

____

161. Explain how to avoid overprocessing and underprocessing.

____

162. Name three areas where test curls should be taken.

1. ______________________ 2. ______________________

3. ______________________

163. Describe the force and temperature of water needed to rinse waving lotion from the hair.

____

164. Name the amount of time required to rinse the hair. ______________________

165. Describe where in the hair structure the waving lotion is rinsed from.

____

166. Identify the area of the head that is more difficult to rinse. ______________________

167. Name two types of hair that require maximum rinsing time.

1. ______________________ 2. ______________________

168. Describe what happens when waving lotion is left in the hair and interferes with the action of the neutralizer.

____

____

169. Explain how insufficient rinsing can cause the hair color to lighten.

____

170. Describe how insufficient rinsing can cause residual perm odors.

____

____

171. Explain why careful blotting is necessary before neutralization.

____

172. Describe how to obtain the best results from towel blotting.

____

173. Explain why rods should not be rocked or rolled while blotting.

____

174. Explain the necessity of removing as much excess water as possible.

____

____

175. Explain when to place a fresh, clean band of cotton around the hairline.

____

NEUTRALIZING

176. Describe how to apply neutralizer to rods.

____

____

177. Specify the amount of time for the neutralizer to remain on the hair. _______________

178. Describe how to remove the rods. _______________

179. Explain what to do with the remaining neutralizer.

____

180. Specify the water temperature needed to rinse the hair. _______________

181. Explain why pulling or using intense heat on freshly permed hair is not recommended.

____

182. Identify the number of hours to wait before hair can be shampooed, conditioned, or treated harshly. _______________

TEN POINTERS FOR A PERFECT PERM

183. List ten pointers for a perfect perm.

    1. _______________
    2. _______________
    3. _______________
    4. _______________

       ____
    5. _______________
    6. _______________
    7. _______________
    8. _______________
    9. _______________
    10. _______________

        ____

184. List the five steps required for final cleanup.

    1. _______________
    2. _______________
    3. _______________
    4. _______________
    5. _______________

DIRECTIONAL WRAPPING

185. Describe directional wrapping.

____

____

186. Name the number of basic directions in which hair can be wrapped. __________

187. List two advantages of directional wrapping.

    1. ____________________ 2. ____________________

188. List the three steps required for directional wrapping.

    1. ____

       ____

    2. ____

    3. ____

BODY WAVES

189. Define body wave. ______________________________

190. Explain why large or extra large rods are used. __________________

191. Explain the similarities between perms and body waves.

____

192. Differentiate between perms and body waves.

____

197. Explain why extra large partings should not be used.

____

194. Explain why the processing time should not be reduced.

____

195. Explain why straight rods, rather than concave, should be used.

____

196. Explain why body waves are softer and wider than regular perm curls.

____

____

PARTIAL PERMING

197. Describe partial perming. ______

198. List three situations for partial perming.
    1. ______
    2. ______
    3. ______

199. Explain how to blend permed hair into the unpermed hair.

    ______

200. Describe where to place a coil of cotton after wrapping.

    ______

201. Explain how to protect the sections that will not be permed from the effects of the waving lotion. ______

202. Describe the effect of waving lotion on unwrapped hair.

    ______

PERMS FOR MEN

203. Name three ways that perms help overcome common hair problems for men.
    1. ______
    2. ______
    3. ______

204. Name the technique most often used for male clients. ______

HEATED CLAMP METHOD

205. Describe what is done after hair is wound on rods and thoroughly saturated with waving lotion. ______

206. Specify when processing begins. ______

207. Describe what happens after the hair has been processed.

    ______

208. List three special control features of the heated clamp method.
    1. ______
    2. ______
    3. ______

SPECIAL SITUATIONS

209. Explain what to do for hair that is excessively dry, brittle, or over-porous.

____

210. Explain what to do if the condition improves.

____

211. Explain why hair previously treated with a sodium hydroxide or "no lye" relaxer should not be permed.

____

____

212. Describe the type of perm to use on tinted, bleached, highlighted/frosted, or previously permed hair in good condition. _______________________

213. Explain when tinted hair should be treated as bleached hair.

____

214. Explain why hair treated with semi-permanent hair color is frequently more resistant to perming.

____

____

215. Specify the length of time to wait before reapplying a semi-permanent color.

____

216. Explain the effects on hair of products containing metallic salts.

____

217. Describe what to mix for a 1–20 test.

____

218. Describe what happens if there are no metallic salts present.

____

219. Describe what happens if the hair contains lead.

____

220. Describe what happens if the hair contains silver.

____

221. Describe what happens if the hair contains copper.

____

____

222. Explain how to tell if hair is no longer coated with metallic salts. _______________

223. Explain the purpose of perming unmanageable, naturally curly hair with an uneven curl pattern. _______________________________

224. Explain why the perm formula must be chosen carefully for naturally curly hair.

____

225. Identify how often clients can return for other perms. ____________________

226. List three factors that determine how often clients can get repermed.

1. ____________________ 2. ____________________

3. ____________________

227. Name three facts that must be noted on the client's record card.

1. ____________________ 2. ____________________

3. ____________________

WORD REVIEW

base	blocking	body wave
concave	density	dexterity
diameter	directional wrapping	elasticity
end wraps	fishhook	neutralizer
overprocessed	partial perm	partings
piggyback	porosity	preliminary
processing time	relaxation	sectioning
texture	underprocessed	wrapping

MATCHING TEST

Insert the correct term or phrase in front of each definition.

base	body wave	density
elasticity	overprocessing	partial perm
partings	porosity	reapplication
sectioning	texture	underprocessing

1. ____________________ the number of hairs per square inch
2. ____________________ the dividing of hair into uniform, working areas
3. ____________________ a condition caused by insufficient processing time of the waving lotion
4. ____________________ the hair's capacity to absorb liquids
5. ____________________ a perm that gives support, not definite curl
6. ____________________ the overall plan for rod placement
7. ____________________ the diameter of each individual hair
8. ____________________ a condition caused by leaving waving lotion on hair too long
9. ____________________ a perm done only on a section of a whole head of hair
10. ____________________ the ability of hair to stretch and contract

RAPID REVIEW TEST

Place the correct word in the space provided in each sentence below.

after	analysis	band
before	bulkiness	coatings
contract	curly	directions
elastic	expand	fishhook
frizzy	large	less
longer	milder	more
neutralize	neutralizer	opposite
penetration	poor	porous
resistant	rod	same
saved	scalp	sensitive
small	stronger	temperature
unevenly	weak	

1. The hair must be free of all ______________ before beginning any perm.
2. The more ______________ the hair, the less processing time it takes.
3. ______________ of hair on the rod prevents penetration of the waving lotion and neutralizer.
4. Opened, unused waving lotion or neutralizer should not be ______________.
5. Perming techniques are basically the ______________ for both male and female clients.
6. Underprocessed hair has a ______________ wave formation.
7. The size of the ______________ controls the size of the curl.
8. Product ______________ must be read and followed carefully.
9. An incorrect pre-perm ______________ can result in poor curl development or hair damage.
10. To prevent ______________ ends, the first turn on the rod should be the end wraps without any of the hair ends between them.
11. The piggyback wrap permits maximum control of the size and tightness of the curl from the ______________ to ends.
12. Hair treated with a semi-permanent hair color is frequently ______________ to perming.
13. The more porous the hair, the ______________ the waving solution required.

14. The wave ridge in fine, thin hair may be __________defined and ___________difficult to read.

15. With the heated clamp method, the ________________of the rods is strictly controlled.

16. Hair should be thinned or texturized ___________the perm.

17. Careful blotting after rinsing assures that the ________________will penetrate the hair completely.

18. To prevent breakage, the ___________should not press into the hair near the scalp.

19. Hair completely lacking in elasticity has lost its ability to ______________after stretching.

20. Vigorous brushing, combing, pulling, or rubbing can cause the scalp to become ______________to perm solutions.

21. Fine hair has a ___________diameter; coarse hair has a ___________diameter.

22. Hair that is tightly wrapped prevents ________________of the waving lotion and neutralizer.

23. Overprocessed hair is very ___________when wet, but ____________when dry.

24. The length of the blocking should never be ____________than the length of the rod.

25. A pre-wrap lotion is recommended for hair that is ______________porous.

MULTIPLE CHOICE TEST

Read each statement carefully, then write the letter representing the word or phrase that correctly completes the statement on the blank line to the right.

1. The degree to which hair absorbs the waving lotion is related to its
   a) texture b) length
   c) elasticity d) porosity ________

2. After perming, hair should not be shampooed, conditioned, or treated harshly for
   a) 24 hours b) 48 hours
   c) 72 hours d) 96 hours ________

3. Alkaline perms are recommended for hair that is
   a) fine and resistant b) tinted and porous
   c) normal and delicate d) bleached ________

4. To create bangs on the forehead, the first two top front curls are wrapped
   a) backward b) forward
   c) toward the right side d) toward the left side ________

5. The determining factor when choosing rod size is the hair
   a) porosity b) density
   c) elasticity d) texture ________

6. Hair containing copper is detected by
   a) a slight lightening b) a rapid lightening
   c) an unpleasant odor d) no reaction ________

7. A method of wrapping that is especially suitable for extra-long hair is the
   a) single halo b) double halo
   c) piggyback d) dropped crown ________

8. Waving lotion that comes in contact with eyes or skin must be
   a) blotted b) wiped
   c) rubbed d) rinsed thoroughly ________

9. Hair must be wrapped smoothly and neatly on each perm rod without
   a) blocking it first b) combing it first
   c) end papers d) stretching ________

10. To check a test curl, unwind a curl about
    a) 1 turn b) 1 1/2 turns
    c) 2 turns d) 2 1/2 turns ________

11. For perms requiring dryer heat, the dryer must be
    a) pre-heated b) warm
    c) lukewarm d) cool ________

12. Stretching or pulling the hair toward the rod while wrapping can cause hair to
    a) discolor b) break
    c) become porous d) grow faster ________

13. A perm rod whose diameter is the same throughout its length is

a) short
b) long
c) concave
d) straight ________

14. On the same day, a client should not receive a perm and a

a) hairstyle
b) hair color
c) manicure
d) pedicure ________

15. When perming hair longer than 6", it is necessary to use

a) small partings
b) average partings
c) large partings
d) extra-large partings ________

16. When the strand is held in an upward position, the curl will rest

a) on-base
b) half on-base
c) off-base
d) half off-base ________

17. Plastic caps must be airtight and cover

a) the rods on top
b) the rods in back
c) the rods on the sides
d) all the rods ________

18. Before starting a perm, hair is shampooed and

a) brushed
b) pulled
c) towel dried
d) oiled ________

19. The action of the waving lotion

a) shrinks the hair
b) expands the hair
c) conditions the hair
d) colors the hair ________

20. Underprocessed hair can be re-permed with a waving lotion that is

a) milder
b) stronger
c) heavier
d) creamier ________

21. A tighter curl at the hair ends with a looser curl at the scalp results from wrapping with

a) short rods
b) long rods
c) concave rods
d) straight rods ________

22. If hair breaks under very slight strain, it has

a) outstanding elasticity
b) very good elasticity
c) good elasticity
d) little or no elasticity ________

23. Hair previously treated with a sodium hydroxide or "no lye" relaxer should not be

a) shampooed
b) combed
c) styled
d) permed ________

24. The size of the partings is determined by hair

a) porosity
b) density
c) elasticity
d) texture ________

25. The double halo wrap is usually used for

a) smaller-size heads
b) average-size heads
c) larger-size heads
d) a smooth crown effect ________

26. During the processing time, optimum curl development occurs

a) only once
b) twice
c) three times
d) four times
____

27. Porosity is easily detected in hair that

a) tangles easily
b) does not tangle
c) is shiny
d) feels smooth
____

28. If hair becomes dry while wrapping, it should be misted lightly with

a) hairspray
b) water
c) styling lotion
d) lubricant
____

29. A client whose perm processes at room temperature should not sit

a) on a chair
b) on a stool
c) on a bench
d) in a draft
____

30. The hair length considered ideal for perming is

a) 1" to 5"
b) 2" to 6"
c) 3" to 7"
d) 4" to 8"
____

31. Severe discoloration or hair damage can result when attempting to perm hair treated with products containing

a) hydrogen peroxide
b) ammonia
c) metallic salts
d) highlighting shampoo
____

32. The best way to blend permed with unpermed hair is to use rods that are

a) one size smaller
b) two sizes smaller
c) one size larger
d) two sizes larger
____

33. For optimum bonding, most neutralizers remain in the hair for

a) 3 minutes
b) 5 minutes
c) 10 minutes
d) 15 minutes
____

34. The main difference between a perm and a body wave is the

a) size of the rods
b) strength of the lotion
c) processing time
d) neutralizing time
____

35. End wraps minimize the danger of hair

a) expansion
b) shrinkage
c) discoloration
d) breakage
____

36. Unmanageable, naturally curly hair with an uneven curl pattern can be permed to form

a) smaller curls
b) tighter curls
c) larger curls
d) frizzier curls
____

37. For wrapping, hair is shampooed and left

a) dripping wet
b) saturated
c) moist
d) dry
____

38. Hair with the cuticle layer lying close to the hair shaft

a) has poor porosity
b) has good porosity
c) is porous
d) is over-porous
____

39. Perms for men can help make sparse hair look

a) thinner
b) fuller
c) darker
d) lighter ________

40. Incomplete rinsing of waving lotion from processed hair can result in

a) excessive curl
b) a fragrant odor
c) darkening of hair color
d) early curl relaxation ________

41. Perm rods are typically made of

a) fiberglass
b) plastic
c) metal
d) acrylic ________

42. With the heated clamp method, processing begins as soon as the

a) rods are wound on hair
b) neutralization begins
c) pre-heated clamps are applied
d) waving lotion is rinsed off ________

43. A coating on the hair could be the result of improper

a) rinsing
b) cutting
c) styling
d) scalp manipulation ________

44. Hair with coarse texture and good elasticity requires

a) small partings and rods
b) large partings and rods
c) small partings, large rods
d) large partings, small rods ________

45. For close-to-the-head hairstyles not requiring fullness or height, rods are wrapped

a) on-base
b) off-base
c) half on-base
d) half off-base ________

46. To help determine in advance how a client's hair will react to a perm, it is necessary to do a

a) patch test
b) strand test
c) predisposition test
d) preliminary test curl ________

47. Hair that feels harsh after being dried and whose elasticity has been excessively damaged is

a) underprocessed
b) overprocessed
c) in good condition
d) suitable for re-perming ________

48. Excess water left in the hair can dilute the neutralizer and cause curls to be

a) weak
b) firm
c) springy
d) bouncy ________

49. The appropriate perm product must be chosen to suit the client's

a) face shape
b) head shape
c) hair type
d) body structure ________

50. The information gathered during the client consultation should be

a) disregarded
b) partially disregarded
c) included on the client's record card
d) taken with a grain of salt ________

Also see *Milady's Standard Theory Workbook.*

Date ________________

Rating ______________

Text Pages 225–280

HAIR COLORING

PREPARATION FOR HAIR COLORING

1. Specify where to record information about the consultation. ____________________

2. Describe the type of lighting necessary for the consultation.

   ____

   ____

3. List two considerations to discuss with the client.

   1. ____________________
   2. ____________________

4. Name two products necessary for clients' at-home hair care.

   1. ____________________ 2. ____________________

PREDISPOSITION TEST

5. Explain when the patch test must be given.

   ____

6. Describe the tint formula used for the patch test.

   ____

7. List the eight steps required for giving a patch test.

   1. ____________________
   2. ____________________
   3. ____________________
   4. ____________________
   5. ____________________
   6. ____________________
   7. ____________________
   8. ____________________

8. Describe a negative skin test. ____________________

9. List five symptoms of a positive skin test.

1. ______________________ 2. ______________________

3. ______________________ 4. ______________________

5. ______________________

10. Explain two consequences of applying an aniline derivative tint to an allergic client.

1. ______________________ 2. ______________________

EXAMINING SCALP AND HAIR

11. List four needs that may be revealed by an examination.

1. ______________________

2. ______________________

3. ______________________

4. ______________________

12. List four conditions that prohibit using an aniline derivative tint.

1. ______________________ 2. ______________________

3. ______________________ 4. ______________________

13. Specify where to record information about a client's hair condition.

____

FORMULATING COLORS

14. Specify what cosmetologists must consider as they begin to formulate hair colors.

____

15. Specify what most hair colors generally contain.

____

16. Explain how color manufacturers classify oxidation tints.

____

17. Explain how cosmetologists can identify the level and base color of the client's hair.

____

18. Identify three predominant base colors of cool colors.

1. ______________________ 2. ______________________

3. ______________________

19. Identify three predominant base colors of warm colors.

1. ______________________ 2. ______________________

3. ______________________

20. Explain the range in tones for formulations of all colors.

____

BASIC RULES FOR COLOR SELECTION

21. Describe how the client's hair must be for color selection. ______________________

22. Explain how to see depth as well as highlights in the hair.

____

____

23. Explain why it is necessary to analyze the depth present in the hair and the depth of the desired color.

____

____

24. Explain why it is necessary to determine the natural highlights as well as the desired highlights.

____

____

25. Explain why it is necessary to know the properties of the product.

____

26. Explain why it is necessary to analyze the condition of the hair.

____

____

FOLLOWING A WORKING PLAN IN TINTING

27. List three results of not following a definite tinting procedure.

    1. ______________________ 2. ______________________

    3. ______________________

28. List two important components of a working plan.

    1. ______________________

    2. ______________________

KEEPING HAIR COLOR RECORDS

29. Explain why keeping accurate records is important.

____

30. List four pieces of information that should be included in a complete record card.

    1. ______________________ 2. ______________________

    3. ______________________ 4. ______________________

STRAND TEST TO CONFIRM COLOR SELECTION

31. List four points of information indicated by a preliminary strand test.
    1. ______
    2. ______
    3. ______
    4. ______

32. List the five steps required to do a strand test.
    1. ______
    2. ______
    3. ______
    4. ______
       ______
    5. ______

RELEASE STATEMENT

33. Specify when to use a release statement. ______

34. Explain the purpose of the release statement.

    ______
    ______
    ______

TEMPORARY COLORING

35. Explain why temporary color lasts only from shampoo to shampoo.

    ______

36. Explain what causes some temporary color to last much longer.

    ______

37. Describe the contents of temporary colors.

    ______

38. List six advantages of temporary colors.
    1. ______
    2. ______
    3. ______
    4. ______
    5. ______
    6. ______

39. List five disadvantages of temporary colors.

1. ______________________________
2. ______________________________
3. ______________________________
4. ______________________________
5. ______________________________

METHODS OF APPLICATION

40. List the implements and materials needed to apply a temporary color.

1. ______________ 2. ______________
3. ______________ 4. ______________
5. ______________ 6. ______________
7. ______________ 8. ______________
9. ______________

41. Describe what is done to the hair before applying a temporary color.

____

42. Explain why the client must be protected with the neck strip and cape.

____

43. List the six steps required to apply a temporary color.

1. ______________________________
2. ______________________________
3. ______________________________
4. ______________________________
5. ______________________________
6. ______________________________

44. List the six steps required in the cleanup.

1. ______________________________
2. ______________________________
3. ______________________________
4. ______________________________
5. ______________________________
6. ______________________________

45. Name four other available forms of temporary color.

1. ______________ 2. ______________
3. ______________ 4. ______________

SEMI-PERMANENT COLORING

46. Describe how semi-permanent color is formulated.

____

47. List three things that semi-permanent color can do.

1. ____________________ 2. ____________________

3. ____________________

48. List one thing that semi-permanent color cannot do. ____________________

49. Explain why clients can change the color at any time or discontinue the effect.

____

50. Name four available forms of semi-permanent color.

1. ____________________ 2. ____________________

3. ____________________ 4. ____________________

51. List four factors upon which color results depend.

1. ____________________ 2. ____________________

3. ____________________ 4. ____________________

52. Identify the length of time semi-permanent colors are formulated to last.

____

53. Explain why the color gradually fades with each shampoo.

____

54. Name two instances when semi-permanent color can be more permanent.

1. ____________________

2. ____________________

ADVANTAGES

55. List five advantages of semi-permanent color.

1. ____________________

2. ____________________

3. ____________________

4. ____________________

5. ____________________

56. Identify a type of semi-permanent color that requires a patch test.

____

TYPES

57. Describe four types of semi-permanent color.

    1. ______________________________
    2. ______________________________
    3. ______________________________
    4. ______________________________

58. List the implements and materials needed to apply a semi-permanent color.

1. ______________	2. ______________
3. ______________	4. ______________
5. ______________	6. ______________
7. ______________	8. ______________
9. ______________	10. ______________
11. ______________	12. ______________
13. ______________	14. ______________
15. ______________	16. ______________

59. List the eight preliminary steps required to apply a semi-permanent color.

    1. ______________________________
    2. ______________________________
    3. ______________________________
    4. ______________________________
       ____
    5. ______________________________
    6. ______________________________
    7. ______________________________
    8. ______________________________

60. List the twelve steps required to apply a semi-permanent color.

1. ______________________________
2. ______________________________
3. ______________________________
4. ______________________________
   ____
5. ______________________________
6. ______________________________
7. ______________________________
8. ______________________________
9. ______________________________
   ____
10. ______________________________
11. ______________________________
12. ______________________________

61. List the six steps required in the cleanup.

1. ______________________________
2. ______________________________
3. ______________________________
4. ______________________________
5. ______________________________
6. ______________________________

PERMANENT HAIR COLORING

62. Describe how far permanent tints penetrate into the hair.

____

____

APPLICATION CLASSIFICATIONS

63. Name the two classifications of permanent hair color.

1. ______________________ 2. ______________________

64. Describe how the desired color is achieved with single-process coloring.

____

65. List two other names for single-process coloring.

1. ______________________ 2. ______________________

66. Give two examples of single-process coloring.

1. ______________________ 2. ______________________

67. Describe how the desired color is achieved with double-process coloring.

____

68. List two other names for double-process coloring.

1. ______________________ 2. ______________________

69. Give two examples of double-process coloring.

1. ____
2. ____

70. Explain when a predisposition test must be given.

____

71. Explain how the client's clothes are protected. ______________________

72. Explain how cosmetologists can protect themselves from allergic reactions.

____

SINGLE-PROCESS TINTS

73. Describe the ability of single-process tints.

____

74. List the main ingredients contained in single-process tints.

1. ______________________ 2. ______________________
3. ______________________ 4. ______________________

75. Explain how color results are altered.

____

76. List five advantages of single-process tints.

1. ____
2. ____
3. ____
4. ____
5. ____

SINGLE-PROCESS TINT FOR LIGHTENING VIRGIN HAIR

77. Define virgin hair.

____

____

78. List the implements and materials needed to apply a single-process tint to lighten virgin hair.

1. ______________ 2. ______________
3. ______________ 4. ______________
5. ______________ 6. ______________
7. ______________ 8. ______________
9. ______________ 10. ______________
11. ______________ 12. ______________
13. ______________ 14. ______________
15. ______________

79. List the eight preliminary steps required to apply a single-process tint.

1. ______________
2. ______________

____
3. ______________
4.

____

____
5. ______________
6. ______________
7. ______________
8. ______________

80. Describe how to section the hair. ______________

81. Explain where to begin applying the tint.

____

82. Identify the size of the subsections. ______________

83. Explain where on the hair strand to apply tint. ______________

84. Give two reasons why hair at the scalp processes faster.

1. ______________ 2. ______________

85. Explain how to check for color development. ______________________

86. Explain where to apply tint after the color has developed on the cold strand.

____

87. Explain how to lather and rinse the color thoroughly.

____

88. Describe how to remove stains from around the hairline.

____

____

89. Describe the type of shampoo used to shampoo hair.

____

90. Name three reasons to apply an acid or a finishing rinse.

1. ______________________ 2. ______________________

3. ______________________

91. Name the final step in the procedure to apply a single-process tint.

____

92. List the six steps required in the cleanup.

1. ______________________
2. ______________________
3. ______________________
4. ______________________
5. ______________________
6. ______________________

SINGLE-PROCESS TINT FOR DARKENING VIRGIN HAIR

93. Explain where to begin application of tint.

____

____

94. Explain where on the hair strand to apply tint. ______________________

95. Explain how to ensure complete coverage of tint.

____

____

SINGLE-PROCESS TINT RETOUCH

96. List the seven preliminary steps required to apply a single-process tint retouch.
    1. ______
    2. ______
    3. ______
    ______
    4. ______
    ______
    5. ______
    6. ______
    7. ______
97. Explain how to section hair for a tint retouch. ______
98. Explain where on the head to begin applying tint.
    ______
99. Specify the size of the subsections. ______
100. Describe where on the hair strand to apply tint. ______
101. Explain why overlapping of color must be avoided.
    ______
102. Name the purpose of strand testing. ______
103. Name three ways to dilute remaining tint mixture.
    1. ______ 2. ______
    3. ______
104. Explain the reason for diluting tint mixture. ______
105. Explain how to lather and rinse color thoroughly.
    ______
106. Describe how to remove stains from around hairline.
    ______
107. Describe the type of shampoo used to shampoo hair.
    ______
108. Name three reasons to apply an acid or a finishing rinse.
    1. ______ 2. ______
    3. ______
109. List the final two steps in the procedure of applying a single-process tint retouch.
    1. ______ 2. ______

HIGHLIGHTING SHAMPOO COLOR

110. List the three ingredients in highlighting shampoo tints.

1. ______________________ 2. ______________________
3. ______________________

111. Name two reasons to use highlighting shampoo colors.

1. ______________________
2. ______________________

112. Explain what is required before applying a highlighting shampoo color. ______________

113. List the two ingredients in highlighting shampoos.

1. ______________________ 2. ______________________

114. Name the reason for using a highlighting shampoo.

____

115. Identify the area of the salon where highlighting shampoo colors are applied.

____

116. Explain how to apply a highlighting shampoo color.

____

117. Specify the length of time to process a highlighting shampoo color.

____

DOUBLE-PROCESS TINT

118. Explain when to pre-lighten a client's hair.

____

119. Describe how to apply a pre-lightener.

____

120. Explain how to remove pre-lightener when the desired shade is reached.

____

121. Explain the purpose of pre-softening a client's hair.

____

122. Describe how to apply a pre-softener. ______________________

123. Explain what a pre-softener does to the hair.

____

124. Describe how to remove pre-softener from the hair. ______________________________

SAFETY PRECAUTIONS

125. List the safety precautions pertaining to hair coloring.

1. ____
2. ____
3. ____
4. ____
5. ____
6. ____
7. ____
8. ____
9. ____
10. ____
11. ____
12. ____
13. ____
14. ____
15. ____
16. ____
17. ____

WORD REVIEW

alkalize
certified
cuticle
FDA
hydrogen peroxide
overlap
permanent
pre-lighten
retouch
single-process coloring
translucent
aniline derivative
consultation
demarcation
filler generator
keratinization
oxidation
pH
preliminary
self-penetrating
single-process tints
virgin
catalyst
cortex
double-process coloring
highlighting
malpractice
patch
predisposition
pre-soften
semi-permanent
temporary
volume

LIGHTENING VIRGIN HAIR

126. List three purposes of the preliminary strand test before lightening.

    1. ______
    2. ______
    3. ______

127. Explain what to do if strand test shows that hair is not light enough.

    1. ______ 2. ______

128. Explain what to do if strand is too light.

    1. ______ 2. ______

129. List three reactions to observe in test strand.

    1. ______ 2. ______
    3. ______

130. Specify when to take a patch test.

    ______

131. List the implements and materials needed for lightening virgin hair.

    1. ______ 2. ______
    3. ______ 4. ______
    5. ______ 6. ______
    7. ______ 8. ______
    9. ______ 10. ______
    11. ______ 12. ______
    13. ______ 14. ______
    15. ______

132. Describe how to protect client's clothing. ______

133. Explain when not to perform the service.

    ______

134. Specify what to check before proceeding with lightener and toner. ______

135. Describe how to section hair. ______

136. Explain where to apply protective cream. ______

137. Explain why lightening formula must be used immediately.

    ______

138. Name the area of the head on which to begin applying lightener.

    ______

139. Identify the size of the partings needed to apply lightener. ______

140. Specify how far from scalp to apply lightener. ______

141. Describe where on the subsection to apply lightener.

___

142. Explain what to do when double-checking application.

___

143. Name two ways to keep lightener moist during development.
    1. ___
    2. ___

144. Explain when to make the first check for lightening action.

___

___

145. Describe how to test for lightening action.

___

146. Identify the size of the partings needed to apply lightener to hair near scalp. ___

147. Describe the temperature of water needed to rinse lightener from hair. ___

148. Name the type of shampoo needed to remove lightener from hair. ___

149. Explain how to avoid tangling the hair. ___

150. Describe how to neutralize the alkalinity of the hair. ___

151. Explain how to dry hair before applying toner.

___

___

152. Explain what to observe in scalp and hair before applying toner.

___

LIGHTENER RETOUCH

153. Describe the purpose of a lightener retouch.

___

154. List three instances when lightener is applied to areas of the hair in addition to the new growth.
    1. ___
    2. ___
    3. ___

155. Specify the area that must be lightened first. ___

156. Specify the length of time to leave lightener mixture on the hair shaft.

___

157. Name two main reasons to consult the client's record card.

1. ______________________________

2. ______________________________

158. Compare the procedure for a lightener retouch to that for lightening a virgin head of hair.

____

____

159. List two reasons for using a cream lightener for a retouch.

1. ______________________________

2. ______________________________

160. Name two results of overlapping.

1. ________________ 2. ________________

161. Draw a line from each illustration to the correct description.

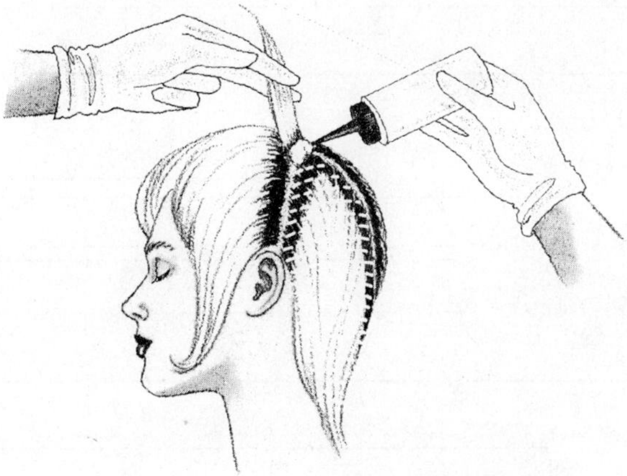

strand testing

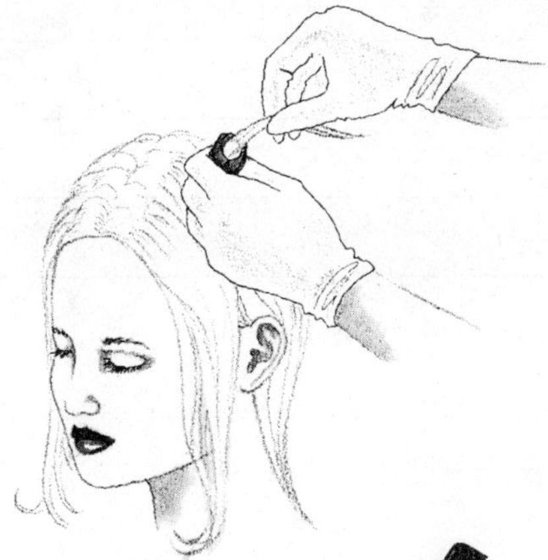

check for complete coverage

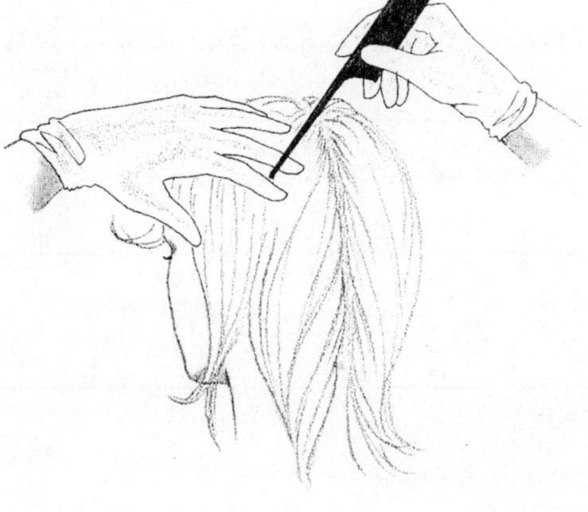

apply lightener to new growth

SAFETY PRECAUTIONS

162. Specify when to give a patch test.

____

163. Explain when to read manufacturer's directions. ______________________________

164. Specify when cosmetologists should wash their hands.

____

165. Explain how to protect the client's clothing. ______________________________

166. Explain when not to apply lightener. ______________________________

167. Describe what to avoid if a shampoo is required before applying a lightener.

____

168. Identify the type of treatment that may be necessary before applying a lightener.

____

169. Specify what the cosmetologist must wear. ______________________________

170. Name two items that must be sanitized prior to use.

    1. ______________________________ 2. ______________________________

171. Name the type of test given prior to applying a lightener. ______________________________

172. Explain why cream lightener should be the thickness of whipped cream.

____

173. Specify how soon to use lightener after it is mixed. ______________________________

174. Explain what to do with leftover lightener. ______________________________

175. Name the areas where lightener is applied first. ______________________________

176. Specify the size of the partings used. ______________________________

177. Explain how to apply lightener for even lightening. ______________________________

178. Explain what the cosmetologist must do until the desired stage is reached.

____

179. Explain how to remove lightener from skin and scalp.

____

180. Specify the maximum length of time to safely leave lightener on scalp.

____

181. Explain what to do if towel around client's neck becomes saturated.

____

182. Describe the shampoo and the water temperature needed to remove lightener.

____

183. Explain how to avoid contamination of lightener products.

____

184. Name the final step in the lightener retouch procedure.

____

SPOT LIGHTENING

185. List three results of careless lightener application.

1. ______________ 2. ______________

3. ______________

186. List the four steps necessary to correct streaked hair.

1. ______________

2. ______________

3. ______________

4. ______________

TONERS

187. Explain why toners require a patch test. ______________

188. Describe toner colors. ______________

189. Name the two parts of a double-process application.

First process: ______________

Second process: ______________

190. Explain what is meant by "foundation."

____

____

191. Explain how cosmetologists can obtain information regarding the correct foundation.

____

____

192. List two reasons to follow the manufacturer's guide closely.

1. ______________

____

2. ______________

____

193. Explain the reason for referring to the laws of color when selecting a toner.

____

194. Identify what must be achieved for toner to develop. ______________

195. List three hair types that may reach the correct color foundation without achieving sufficient porosity.

1. ____________________ 2. ____________________

3. ____________________

196. Specify when to give a patch test. ____________________

197. Explain when to do the strand test to save time. ____________________

198. Name two requirements that must be met to proceed with the application.

1. ____________________ 2. ____________________

199. List the implements and materials needed for a toner application.

1. ____________________ 2. ____________________

3. ____________________ 4. ____________________

5. ____________________ 6. ____________________

7. ____________________ 8. ____________________

9. ____________________ 10. ____________________

11. ____________________ 12. ____________________

13. ____________________ 14. ____________________

15. ____________________

200. List the steps required in the preparation for a toner application.

1. ____________________

2. ____________________

3. ____________________

4. ____________________

5. ____________________

6. ____________________

7. ____________________

8. ____________________

9. ____________________

10. ____________________

____

201. Specify when cosmetologists should wear gloves. ____________________

202. Name two factors that are crucial for good color results.

1. ____________________ 2. ____________________

203. Identify two types of toners whose applications may vary.

1. ____________________ 2. ____________________

204. Name the number of sections to divide hair. ____________________

205. Describe what to avoid while sectioning the hair. ____________________

206. Specify the size of the partings. ____________________

207. Describe where on the hair strand to apply toner. ____________________

208. Explain when to work toner through ends.

____

209. Explain what to do if ends tend to absorb too much color.

____

____

210. Name the reason for leaving hair loose after applying additional mixture on hair.

____

211. Explain how to determine the timing of the toner.

____

212. Describe how to remove the toner. ____________________

213. Describe how the hair is shampooed and rinsed. ____________________

214. List three reasons for applying a finishing rinse.

1. ____________________ 2. ____________________

3. ____________________

215. Name what must be removed from the skin, hairline, and neck. ____________________

216. Name what to avoid when styling the hair. ____________________

217. Name the last two steps in the toner application procedure.

1. ____________________ 2. ____________________

218. Identify the stage to pre-lighten new growth for a toner retouch.

____

219. Describe two possible ways to apply the toner.

1. ____________________

2. ____________________

SPECIAL EFFECTS HIGHLIGHTING

220. Explain what is involved in special effects highlighting.

____

221. Describe three effects of using light colors.

1. ____________________

2. ____________________

3. ____________________

222. Describe three effects of using dark colors.

1. ______________________________

2. ______________________________

3. ______________________________

METHODS FOR HIGHLIGHTING

223. Name the three main methods to highlight hair.

1. ______________ 2. ______________

3. ______________

224. Explain what is involved in the cap technique.

____

225. Describe how to achieve a subtle look.

____

226. Describe how to achieve a more dramatic look.

____

227. Name two types of products used to lighten the hair.

1. ______________ 2. ______________

228. Describe how to remove the lightener.

____

229. Specify what is done to the lightened hair after it is towel-blotted and conditioned.

____

230. Explain what is involved in the foil technique.

____

231. Describe what to do after placing the selected strands over foil or plastic wrap.

____

232. Explain the purpose of folding the foil. ______________

233. Name the advantage of using the foil technique.

____

234. Explain what is involved in freehand techniques.

____

235. List three ways that lightener can be applied.

1. ______________ 2. ______________

3. ______________

236. Explain how freehand techniques can be used effectively.

____

SPECIAL PROBLEMS IN HAIR COLORING

237. List three things the hair colorist must do prior to each hair coloring service.

1. ______________________ 2. ______________________
3. ______________________

238. Describe two reasons why even skilled colorists occasionally have a hair coloring problem.

1. ____

____

2. ____

____

DAMAGED HAIR

239. Name four things that cause damage to hair.

1. ______________________ 2. ______________________
3. ______________________ 4. ______________________

240. Identify three coating compounds that can prevent color penetration.

1. ______________________ 2. ______________________
3. ______________________

241. List six preventative and corrective steps that help prevent hair damage.

1. ____
2. ____
3. ____
4. ____
5. ____
6. ____

242. Identify seven hair conditions considered to be damaged.

1. ______________________ 2. ______________________
3. ______________________ 4. ______________________
5. ______________________ 6. ______________________
7. ______________________

243. Explain when damaged hair should receive reconditioning treatments.

____

____

____

244. Explain why cosmetologists must analyze the hair and consult with the client regarding the source of the hair damage.

____

245. Name three ways to avoid damaging hair during the shampoo.

1. ______
2. ______
3. ______

246. Identify the implement used to apply liquid conditioner and those used to apply cream conditioner.

______

______

247. Name the implement used to blend the conditioner through the hair.

______

248. Specify what is placed over the head before applying heat. ______

249. Explain when to proceed with the coloring service.

______

______

FILLERS

250. Define fillers.

______

______

251. Explain when to use conditioner fillers.

______

252. Explain the advantage of applying the conditioner filler immediately before applying the color.

______

253. Explain when color fillers are recommended.

______

______

254. List seven advantages of using a color filler.

1. ______
2. ______
3. ______
4. ______
5. ______
6. ______
7. ______

255. Describe three ways that color fillers may be used.

1. ______________________________
2. ______________________________
3. ______________________________

256. Explain how to select the color filler to obtain satisfactory results.

____

257. Specify what must be present for natural-looking hair color.

____

TINT REMOVAL

258. Define tint or color removers.

____

259. Describe what tint removers may contain. ______________________________

260. Specify what tint removers are sometimes mixed with. ______________________________

261. List the sixteen steps necessary in the procedure to remove tint.

1. ______________________________
2. ______________________________
3. ______________________________
4. ______________________________
5. ______________________________
6. ______________________________
7. ______________________________
8. ______________________________
9. ______________________________
10. ______________________________
11. ______________________________
12. ______________________________
____
13. ______________________________
14. ______________________________
15. ______________________________
16. ______________________________

262. Explain what to do if hair cannot withstand tinting.

____

TINT BACK TO NATURAL COLOR

263. Describe where to check hair for its natural color. ____________________

264. Describe how a cosmetologist can compromise with the dark color that matches the client's natural hair color.

____

265. Specify where to record all observations and treatments. ____________________

266. Explain why cosmetologists must check the results of the patch test.

____

267. List three things to do to hair before sectioning it.

1. ____________________ 2. ____________________

3. ____________________

268. Specify the number of sections into which to divide hair. ____________________

269. Name what must be replaced by applying a filler. ____________________

270. Explain how to process the filler. ____________________

271. Describe what to do immediately after filler is processed.

____

272. Specify the number of sections into which to divide hair. ____________________

273. Specify the size of the subsections needed to apply color formula. ____________________

274. Describe how to apply the tint.

____

____

275. Explain when to apply tint to porous ends.

____

276. Explain when entire head may be soap capped to blend color.

____

277. Describe the type of shampoo used to remove tint from hair. ____________________

278. Identify what is used to close the cuticle and prevent fading. ____________________

279. Name what is used to replace moisture in the hair. ____________________

280. Name two things to avoid when styling the hair.

1. ____________________ 2. ____________________

281. Explain why clients should be offered the opportunity to purchase high-quality products.

____

282. List the last two steps in the tint-back procedure.

1. ____________________ 2. ____________________

WORD REVIEW

cap technique
deterioration
foil technique
highlighting
primary
retouch
contamination
diffuse
foundation
penetration
reconditioning
toner
demarcation
filler
freehand technique
pre-lighten
regrowth

MATCHING TEST

Insert the correct term or phrase in front of each definition.

aniline derivative tint
overlapping
predisposition test
pre-softening
semi-permanent hair color
tint back
consultation
oxidation
pre-lightening
release statement
single-process coloring
virgin hair
double-process coloring
patch test
preliminary test strand
retouch
temporary hair color

1. ______________________________ a process that achieves the desired results with one application of product
2. ______________________________ a procedure that determines whether a client is allergic to an aniline derivative tint
3. ______________________________ a hair color formulated to fade gradually with each shampoo
4. ______________________________ the application of hair color to the new growth of hair
5. ______________________________ another name for oxidizing penetrating tint
6. ______________________________ a conference between client and cosmetologist to determine a suitable hair color
7. ______________________________ the process of removing color from the hair prior to application of a tint or toner
8. ______________________________ a process requiring two separate applications of products
9. ______________________________ a procedure to determine whether the correct color selection was made
10. ______________________________ a chemical reaction that occurs when an aniline derivative tint is mixed with hydrogen peroxide
11. ______________________________ hair that has not received chemical services
12. ______________________________ another name for patch test
13. ______________________________ a hair color formulated to last until the next shampoo
14. ______________________________ the application of tint or lightener beyond the line of demarcation

15. ______________________ a process that opens the cuticle, making hair more receptive to color

RAPID REVIEW TEST

Place the correct word in the space provided in each sentence below.

base	black	blond
close	consultation	contagious
cortex	cuticle	darker
draping	hairline	hands
highlights	level	lighter
negative	open	overlapped
permanent	positive	record
release statement	retouching	same
sanitized	self	semi-permanent
soften	stain	strip
temporary	thin	

1. Cosmetologists must wash their ____________ before and after serving each client.

2. A ______________ skin test shows no sign of inflammation.

3. When selecting a hair color, it's important to identify the hair's depth as well as its ________________.

4. ________________ hair colors may rub off on pillow or collar.

5. An advantage of single-process tints is that they can produce shades from the deepest ____________ to the lightest ____________.

6. Harsh or alkaline shampoos ___________ hair color.

7. One of the most important steps in the hair coloring service is the __________________.

8. The coating of temporary hair color is ___________ and may not cover hair evenly.

9. Aniline derivative tints should not be applied when ________________ scalp or hair disorders are present.

10. The client's clothing is protected by ______________ with a towel and tint cape.

11. When semi-permanent hair color is applied to extremely porous hair, the results can be more ________________.

12. After shampooing color from the hair, an acid or finishing rinse is applied to ____________ the cuticle.

13. Before formulating a tint for the hair, a color's predominant ____________ and ____________ of lightness or darkness must be identified.

14. Lines of demarcation occur when hair color is __________________ during a retouch application.

15. To repeat successful services and avoid difficulties, cosmetologists must keep an accurate ______________ of each hair color.

16. Semi-permanent hair colors are ____________ -penetrating.

17. All towels, combs, brushes, and applicator bottles used during a tint must be ________________.

18. A temporary hair color applied to porous hair may ____________ it.

19. The tint used for the patch test must be the ____________ formula as that used for the hair coloring service.

20. Protective cream is applied around the _______________ and over the ears.

21. The color molecules in semi-permanent hair coloring penetrate the ______________.

22. A document that dismisses the school or salon owner from responsibility for accidents or damages is the _______________________.

23. A _______________ skin test shows redness, swelling, burning, itching, and blisters.

24. Single-process tints can color the hair ______________ or ______________ than the client's original shade.

25. __________________ is not required with semi-permanent hair colors.

MULTIPLE CHOICE TEST

Read each statement carefully, then write the letter representing the word or phrase that correctly completes the statement on the blank line to the right.

1. A patch test is not required prior to applying

   a) a highlighting shampoo b) a single-process tint
   c) an aniline derivative tint d) a highlighting shampoo tint ________

2. Colors formulated with yellow, orange, or red as their predominant base color are

   a) warm b) cool
   c) neutral d) non-colors ________

3. Applying an aniline derivative tint to a client who shows symptoms of allergy could result in

   a) respect from co-workers b) a malpractice suit
   c) an increase in hair-color clients d) an increase in salary ________

4. Tints are mixed in applicator bottles or bowls made of plastic or

   a) metal b) tin
   c) copper d) glass ________

5. When gray hair is present, the tint application usually begins

   a) in the back b) in the crown
   c) in the front d) on the hair ends ________

6. Temporary hair color coats the hair's

   a) cuticle b) cortex
   c) medulla d) follicle ________

7. Practically all permanent hair coloring is done with oxidizing penetrating tints which contain

   a) certified colors b) compound dyes
   c) metallic dyes d) aniline derivatives ________

8. The hair at the scalp processes faster due to

   a) sebum b) perspiration
   c) body heat d) dead epithelial cells ________

9. During a retouch, overlapping of color can cause hair

   a) breakage b) elasticity
   c) porosity d) retention ________

10. Highlighting shampoos are a mixture of shampoo and

    a) ammonia b) hydrogen peroxide
    c) thioglycolic acid d) aniline derivative tint ________

11. Prior to applying an aniline derivative tint, the hair should not be

    a) analyzed b) brushed
    c) sectioned d) combed ________

12. Hair colors that gradually fade with each shampoo are

    a) temporary b) semi-permanent
    c) permanent d) aniline derivative ________

13. When lightening virgin hair, the tint is applied

a) 1/8" from scalp
b) 1/4" from scalp
c) 1/2" from scalp
d) 1" from scalp ________

14. Most hair colors generally contain a mixture of

a) primary colors
b) secondary colors
c) tertiary colors
d) complementary colors ________

15. For best results when applied to hair, tints must be mixed

a) the day before
b) two hours before
c) an hour before
d) immediately before ________

16. Semi-permanent color cannot

a) add highlights
b) blend gray
c) deepen color tones
d) lighten hair ________

17. Stains around the hairline are removed with shampoo, stain remover, or

a) laundry detergent
b) chlorine bleach
c) scouring powder
d) remaining tint mixture ________

18. Aniline derivative tints should not be used on hair treated with

a) hair spray
b) styling lotion
c) a metallic or compound dye
d) a permanent wave ________

19. Most temporary hair colors are applied to hair after it has been shampooed and

a) left dripping wet
b) towel dried
c) blow dried
d) oiled ________

20. Chemicals in swimming pools can turn blond hair

a) red
b) yellow
c) blue
d) green ________

21. Diluted tint mixture is applied to

a) eyebrows
b) eyelashes
c) faded hair ends
d) hair nearest the scalp ________

22. A client who desires a drastically lighter shade should first have his/her hair

a) shampooed
b) conditioned
c) pre-softened
d) pre-lightened ________

23. Temporary colors last

a) from shampoo to shampoo
b) from 2 to 3 shampoos
c) from 4 to 6 shampoos
d) indefinitely ________

24. Leftover tint mixed with peroxide must be

a) rebottled
b) resealed
c) discarded
d) used on another client ________

25. A release statement is a document signed by a client in the salon before receiving

a) a shampoo
b) a hair style
c) a haircut
d) a chemical service ________

26. Application of single-process tints should begin where hair is most

a) porous
b) curly
c) elastic
d) resistant ________

27. When rinsing color from a client's hair, the water should be

a) hot
b) very warm
c) lukewarm
d) cold
____

28. Semi-permanent colors do not require

a) consulting with client
b) selecting a color
c) mixing with hydrogen peroxide
d) reading manufacturer's directions
____

29. Hair coloring products should not come in contact with the client's

a) scalp
b) eyes
c) hair strand
d) hair ends
____

30. Highlighting shampoo tints are a mixture of shampoo, aniline derivative tint, and

a) water
b) ammonia
c) hydrogen peroxide
d) thioglycolic acid
____

31. Aniline derivative tints enter the hair's

a) cuticle
b) cortex
c) medulla
d) follicle
____

32. Cosmetologists should protect their hands by wearing

a) band aids
b) gloves
c) long fingernails
d) lots of jewelry
____

33. Color products that have been approved by the FDA for use in cosmetics are

a) certified colors
b) complementary colors
c) secondary colors
d) tertiary colors
____

34. Hair that has not been damaged by natural factors such as wind and sun is considered to be

a) sun-streaked hair
b) highlighted hair
c) virgin hair
d) overprocessed hair
____

35. The timing needed to achieve the desired results is determined by the

a) patch test
b) predisposition test
c) preliminary test strand
d) release statement
____

36. Most single-process tints are formulated to be used with

a) 10 volume peroxide
b) 20 volume peroxide
c) 30 volume peroxide
d) 40 volume peroxide
____

37. During the consultation and selection of a hair color, the client's hair must be

a) clean and dry
b) clean and damp
c) dirty and dry
d) dirty and damp
____

38. Aniline derivative tints should not be applied to scalps with

a) a dry condition
b) an oily condition
c) dandruff
d) irritations or eruptions
____

39. When coloring hair, cosmetologists must follow the directions of

a) the client
b) the salon manager
c) the product manufacturer
d) a co-worker
____

40. Temporary color cannot

a) bring out highlights
b) lift color pigment
c) tone down overlightened hair
d) neutralize a yellowish tinge ________

41. In order for gray, resistant hair to readily absorb tint, it should first be

a) brushed
b) shampooed
c) pre-softened
d) pre-lightened ________

42. No patch test is required prior to applying

a) a temporary hair color
b) a single-process tint
c) an aniline derivative tint
d) a highlighting shampoo tint ________

43. When mixed with single-process tints, the hydrogen peroxide is activated by

a) a lightening agent
b) a shampoo
c) a vegetable tint
d) an alkalizing agent ________

44. Hair color should suit the client's

a) spouse
b) best friend
c) wardrobe
d) skin tones ________

45. When tinting virgin hair close to, or darker than, the natural hair color, the tint is applied

a) at the scalp first
b) at the ends first
c) from scalp to ends
d) one-half inch from scalp to ends ________

46. Before applying a hair color, a careful examination of the scalp and hair may indicate the need for

a) a shampoo
b) a new style
c) reconditioning treatments
d) more back-combing ________

47. Prior to any application of aniline derivative color, a patch test must be given at least

a) 4 to 6 hours before
b) 8 to 10 hours before
c) 12 to 16 hours before
d) 24 to 48 hours before ________

48. Semi-permanent colors last

a) from shampoo to shampoo
b) from 2 to 3 shampoos
c) from 4 to 6 shampoos
d) indefinitely ________

49. Colors formulated with blue, green, or violet as their predominant base color are

a) warm
b) cool
c) neutral
d) non-colors ________

50. When done on the client's head, the preliminary strand test is usually given

a) in the crown
b) in the back
c) on the side
d) on the hairline ________

MATCHING TEST

Insert the correct term or phrase in front of each definition.

color filler	conditioner filler	foundation
overlapping	patch test	pre-lighten
preliminary strand test	retouch	spot lightening
tint back	tint remover	toner

1. ______________________ a specialized preparation designed to equalize porosity and deposit a base color in one application

2. ______________________ the correction of a color using a lightener to even out streaks or dark spots

3. ______________________ a procedure that determines whether a client is allergic to an aniline derivative toner

4. ______________________ a commercial product used to remove penetrating tints

5. ______________________ the first step in double-process haircoloring

6. ______________________ a product applied immediately before applying color to recondition damaged hair

7. ______________________ the application of color or lightener beyond the line of demarcation

8. ______________________ the return of hair to its original or natural color

9. ______________________ the color left in the hair after it has gone through the seven stages of lightening

10. ______________________ a procedure that determines processing time needed, condition of hair after lightening, and end results

11. ______________________ a pastel color used on the hair after it has been pre-lightened

12. ______________________ the application of color or lightener to new growth of hair

RAPID REVIEW TEST

Place the correct word in the space provided in each sentence below.

acidity	alkalinity	base
circulation	color	contamination
deterioration	gloves	hair
hands	moist	outgrowth
over	partial	penetration
reconditioning	under	

1. The cosmetologist's hands are protected from the lightener by protective ____________

2. All product bottles and containers should be capped immediately to avoid ____________________

3. Shampooing with hands under __________ helps to avoid tangling.

4. A sign that hair has been ___________ -lightened is when more red, yellow, or orange pigment than desired remain in the hair.

5. The _______________ of the hair is neutralized with an acid or normalizing rinse.

6. For the lightener to continue processing on the hair, it must be kept ____________

7. Coating compounds such as hair sprays, styling agents, and some conditioners can prevent the ________________ of hair color.

8. Cosmetologists must wash their ____________ before and after servicing each client.

9. Damaged hair should receive __________________ treatments prior to and after the application of chemical processes.

10. Leaving hair loose after applying the toner permits better air ________________

11. ___________ -lightened hair "grabs" the base color of the toner.

12. A retouch matches the _______________ to the rest of the lightened hair.

13. Cosmetologists must refer to the laws of ___________ to select a toner that will neutralize or tone the pre-lightened hair to the desired shade.

14. Special effects highlighting involves any technique of ____________ lightening or coloring.

15. After preparing the lightening formula, it must be used immediately to prevent __________________

MULTIPLE CHOICE TEST

Read each statement carefully, then write the letter representing the word or phrase that correctly completes the statement on the blank line to the right.

1. The first process in a double-process application is the
   a) lightener b) toner
   c) semi-permanent hair color d) temporary hair color ________

2. The application of lightener is started in the section where hair is
   a) lightest b) most porous
   c) most resistant d) most fragile ________

3. The cap method of highlighting involves pulling hair through a perforated cap with
   a) a needle b) a hook
   c) an orangewood stick d) a fingernail ________

4. Before the application of a toner, a patch test is given at least
   a) 4 to 6 hours prior b) 10 to 12 hours prior
   c) 16 to 18 hours prior d) 24 to 48 hours prior ________

5. When removing the lightener from the hair, cosmetologists must avoid
   a) using cool water b) using a mild shampoo
   c) removing all traces of lightener d) tangling fragile hair ________

6. The timing needed to achieve the desired results is determined by the
   a) patch test b) release statement
   c) preliminary strand test d) predisposition test ________

7. Prior to the application of a lightener or toner, the hair is divided into
   a) two sections b) three sections
   c) four sections d) six sections ________

8. To obtain satisfactory results, cosmetologists must select the color filler that will replace the missing
   a) primary color b) secondary color
   c) tertiary color d) complementary color ________

9. Colors that cause an area to appear smaller are
   a) light b) dark
   c) primary d) secondary ________

10. Leftover prepared lightener must be
    a) used on another client b) discarded
    c) rebottled d) resealed ________

11. Tint removers are sometimes mixed with
    a) cream lightener b) oil lightener
    c) ammonium thioglycolate d) hydrogen peroxide ________

12. To ensure accurate coverage of the lightener, the size of the hair partings is
    a) 1/8" b) 1/4"
    c) 3/8" d) 1/2" ________

13. Color fillers may be used full strength or diluted with

a) ammonia b) cetyl alcohol
c) a mild shampoo d) distilled water ________

14. Clients should use high-quality products at home to prevent

a) scalp massage b) ease in styling
c) color from being stripped from hair d) hair from being conditioned ________

15. If the towel around the client's neck becomes saturated, it should be removed and replaced to avoid irritation to the client's

a) hair b) skin
c) clothes d) facial features ________

16. Hair that is rough, faded, or brittle is considered to be

a) normal b) healthy
c) resilient d) damaged ________

17. Lightener that accidentally drips on the skin is removed with a

a) cool, damp towel b) warm, damp towel
c) warm, dry towel d) cool, dry towel ________

18. For the toner to develop properly, hair must first achieve sufficient

a) texture b) elasticity
c) porosity d) density ________

19. The application of lightener to virgin hair should begin

a) at the scalp b) 1/8" from scalp
c) 1/4" from scalp d) 1/2" from scalp ________

20. Remaining toner mixture can be diluted with mild shampoo, conditioner, or distilled water before applying to

a) regrowth b) cold strand
c) resistant hair strands d) overporous hair ends ________

21. Checking tinted or lightened hair for its natural color is done

a) next to the scalp b) at the hair ends
c) one inch from the scalp d) one inch from the ends ________

22. Cream lightener should be the consistency of

a) peanut butter b) water
c) whipped cream d) molasses ________

23. An advantage of using a color filler is to help hair to

a) hold color b) fade faster
c) remain streaked d) remain dull ________

24. A method of highlighting that involves weaving, lightening, and folding small strands of hair into a protective wrap is called the

a) cap technique b) foil technique
c) freehand technique d) toning technique ________

25. All observations and treatments regarding the client's hair color service are recorded on the

a) release statement
b) appointment book
c) back of the sales slip
d) client's record card
____

26. Hair lightener must not be applied to clients having
   a) dark-brown hair
   b) light-brown hair
   c) scalp irritation or inflammation
   d) coarse, thick hair
   ____

27. A purpose of the finishing rinse is to close the hair
   a) follicle
   b) cuticle
   c) cortex
   d) medulla
   ____

28. Overlapping of lightener onto previously lightened hair can cause hair
   a) breakage
   b) retention
   c) elasticity
   d) porosity
   ____

29. Toners consist of colors that are
   a) dark and harsh
   b) dark and delicate
   c) pale and delicate
   d) pale and muddy
   ____

30. Prior to applying a lightener, the hair should not be
   a) analyzed
   b) brushed
   c) combed
   d) sectioned
   ____

31. When using any hair coloring product, cosmetologists must follow the directions of the
   a) salon owner
   b) salon manager
   c) client's former hair colorist
   d) product manufacturer
   ____

32. Lightener can be left safely on the scalp area a maximum of
   a) 1/2 hour
   b) 1 hour
   c) 2 hours
   d) 3 hours
   ____

33. Colors that cause an area to appear larger are
   a) light
   b) dark
   c) secondary
   d) tertiary
   ____

34. Before the application of a lightener, protective cream is applied around the hairline and
   a) over the lips
   b) around the eyes
   c) down the nose
   d) over the ears
   ____

Also see *Milady's Standard Theory Workbook.*

Date ________________

Rating ______________

Text Pages 281–300

CHEMICAL HAIR RELAXING AND SOFT CURL PERMANENT

ANALYSIS OF CLIENT'S HAIR

1. Name three methods used to recognize qualities of hair.

   1. ______________________ 2. ______________________
   3. ______________________

2. List four judgments the cosmetologist must make about the hair before attempting to give a relaxing treatment.

   1. ______________________ 2. ______________________
   3. ______________________ 4. ______________________

CLIENT'S HAIR HISTORY

3. Explain the purpose of keeping records of each chemical hair relaxer.

   ____

4. List what to include in the client's records.

   1. ______________________ 2. ______________________
   3. ______________________ 4. ______________________

5. Describe the purpose of the release statement.

   ____

   ____

6. Explain why hair treated with metallic dye must not be chemically relaxed.

   ____

7. Name two steps that determine how the client will react to the relaxer.

   1. ______________________ ______________________

SCALP EXAMINATION

8. Name three conditions that may be detected by a scalp analysis.

   1. ______________________ 2. ______________________
   3. ______________________

9. Explain what happens if the scalp is scratched during the scalp analysis.

____

____

10. Explain what to do if scalp eruptions or abrasions are present.

____

11. Explain what to do if the hair is not in a healthy condition.

____

____

STRAND TESTS

12. Describe the purpose of testing the hair for porosity and elasticity.

____

13. Explain the purpose of the finger test.

____

14. Explain how to do a finger test.

____

____

15. Describe how to determine if hair is porous. ______________________________

16. Explain the purpose of the pull test. ______________________________

17. Explain how to do a pull test.

____

18. Describe how to determine if hair has elasticity.

____

19. Explain the purpose of the relaxer test.

____

20. Explain (briefly) how to do a relaxer test.

____

____

____

____

CHEMICAL HAIR RELAXING PROCESS (WITH SODIUM HYDROXIDE)

21. Identify whose directions to follow when using products containing sodium hydroxide, or any other kind of product. ______________________________

22. List the equipment, implements, and materials required for a sodium hydroxide relaxer.

1.	______________	2.	______________
3.	______________	4.	______________
5.	______________	6.	______________
7.	______________	8.	______________
9.	______________	10.	______________
11.	______________	12.	______________
13.	______________	14.	______________
15.	______________	16.	______________
17.	______________	18.	______________
19.	______________		

PREPARATION

23. List the seven steps required for preparation of a sodium hydroxide relaxer.
    1. ______________________________
    2. ______________________________
    3. ______________________________
       ____
    4. ______________________________
    5. ______________________________
    6. ______________________________
    7. ______________________________

PROCEDURE

24. Specify the number of sections in which to part the hair. ______________

25. Explain what to do if moisture or perspiration is present on the scalp.

    ____

    ____

26. Describe the purpose of the protective base.

    ____

27. Describe where to apply protective base.

    ____

    ____

28. Explain the type of protection needed when using a "no base" relaxer.

____

APPLYING THE CONDITIONER-FILLER

29. List two benefits of the conditioner-filler.
    1. ____
    ____
    2. ____
    ____

30. Describe how to apply conditioner-filler for complete benefits.

____

____

31. Explain why heat should be avoided.

____

32. Explain why cosmetologists must wear protective gloves. ______________________

APPLYING THE RELAXER

33. Identify the number of sections into which to divide the head. ______________________

34. Explain why processing cream is applied last to scalp area and hair ends.

____

____

35. Name three methods used to apply the hair relaxer.
    1. ______________________ 2. ______________________
    3. ______________________

COMB METHOD

36. Name the area of the head on which to begin applying the relaxer.

____

37. Identify the size of the partings. ______________________

38. Name the part of the comb used to apply the relaxer. ______________________

39. Specify the distances from the scalp and from the ends to apply relaxer.

____

40. Explain when to apply relaxer to each side of strand.

____

____

41. Explain what to do when the head is gone over the second time.

____

____

42. Give two reasons for smoothing the cream through the hair.
    1. _______________
    2. _______________

43. Describe an alternate method of applying relaxer.

____

____

BRUSH OR FINGER METHOD

44. Compare the brush or finger method of applying the relaxer to the comb method.

____

____

PERIODIC STRAND TESTING

45. Describe two methods of strand testing while applying relaxer.
    1. _______________
    2. _______________

46. Explain how to tell if strand is sufficiently relaxed. _______________

47. Explain how to tell if strand should continue processing.

____

RINSING OUT THE RELAXER

48. Describe the temperature of the water necessary to rinse relaxer from the hair.

____

49. Explain what happens if the water temperature is too hot.

____

____

50. Explain what happens if the water temperature is too cold.

____

51. Name two reasons for using a direct force of water.
    1. _______________ 2. _______________

52. Explain what happens if relaxer is not completely removed.

____

53. Explain what to do if relaxer or rinse water gets into client's eyes.

____

SHAMPOOING/NEUTRALIZING

54. Describe two methods of neutralizing provided by manufacturers.
    1. ____
    2. ____

55. List two actions to avoid when shampooing the hair.
    1. ______________________ 2. ______________________

56. Describe the position of the fingers while manipulating the shampoo.

____

57. Explain how to neutralize after shampooing.

____

____

58. List three reasons for using the comb.
    1. ____
    2. ____
    3. ____

59. List the seven steps required to complete the relaxer.
    1. ____
    2. ____
    3. ____
    4. ____
    5. ____
    6. ____
    7. ____

    ____

60. Explain why following the manufacturer's directions is always necessary.

____

HOT THERMAL IRONS

61. Name two ways to avoid hair breakage.
    1. ______________________ 2. ______________________

62. Specify the temperature setting necessary when thermal curling chemically relaxed hair.

____

63. Number the illustrations (from 1–5) in the correct order for applying a relaxer.

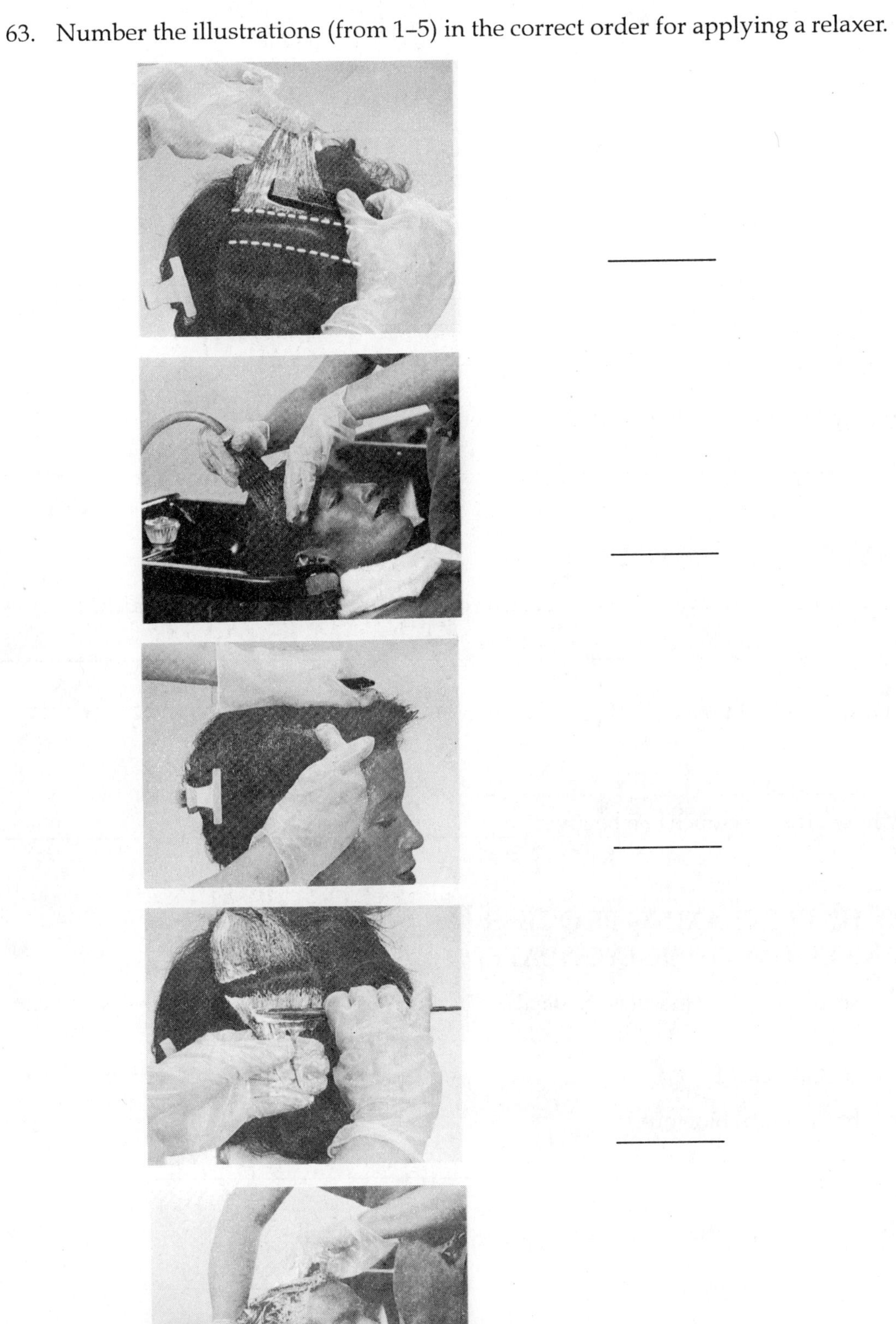

____

____

____

____

____

APPLYING THE CONDITIONER

64. Describe two reasons for applying a conditioner before setting the hair.
    1. ______________________________
    2. ______________________________

65. Differentiate between cream-type conditioners and protein-type (liquid) conditioners.

____

____

____

66. Explain why tension must be avoided while setting the hair.

____

SODIUM HYDROXIDE RETOUCH

67. Name one difference between the regular hair relaxing treatment and the retouch.

____

68. Explain how to avoid breakage of previously treated hair.

69. Specify how often a retouch can be given. ______________________________

CHEMICAL HAIR RELAXING PROCESS (WITH AMMONIUM THIOGLYCOLATE)

First part of this section covered in *Milady's Standard Theory Workbook.*

CHEMICAL BLOWOUT

70. Describe the chemical blowout.

____

71. Name two types of products used for a chemical blowout.

____

72. Explain what happens if the hair is over-relaxed.

____

73. Specify when the hair is shampooed for each type of product.

    Thio: ______________________________

    Sodium hydroxide: ______________________________

EQUIPMENT, IMPLEMENTS, AND MATERIALS

74. List the four items needed for the chemical blowout that are in addition to those needed for a regular chemical relaxing.

1. ______________________ 2. ______________________

3. ______________________ 4. ______________________

PROCEDURE

75. List the seven steps required for a chemical blowout.

1. ______________________
2. ______________________
3. ______________________
4. ______________________

____

5. ______________________
6. ______________________
7. ______________________

76. Describe how to dry the hair for a blowout style.

____

77. Explain how to check the progress of the haircut.

____

REVIEW OF SAFETY PRECAUTIONS

78. Explain what to do if scalp abrasions are present.

____

79. Explain what to do if hair is damaged.

____

80. Identify the type of product that does not require a shampoo prior to its application.

____

81. Identify the type of product that cannot be applied over a thio relaxer.

____

82. Identify the type of product that cannot be applied over a sodium hydroxide relaxer.

____

83. Explain why a strong relaxer should not be used on fine hair. ______________________

84. Explain why excessively hot irons should not be used on chemically relaxed hair.

____

85. Identify the purpose of applying a protective base to the scalp prior to a sodium hydroxide relaxer. ______

86. Identify the most important part of the client's face to protect. ______

87. Name three areas to avoid accidentally spreading the relaxer.

1. ______ 2. ______

3. ______

88. Explain the purpose of frequent strand testing.

______

89. Explain the necessity of a thorough rinsing of relaxer from the hair.

______

90. Describe how to direct the stream of water. ______

91. Describe how long to continue wearing protective gloves.

______

92. Explain why rinsing is done from scalp to hair ends. ______

93. Name two things to avoid when combing the hair after completing the relaxing process.

1. ______ 2. ______

94. Identify what to apply to scalp and hair before setting. ______

95. Describe what to avoid when retouching the new growth.

______

96. Explain what to do if hair is treated with metallic dye.

______

97. Describe what to do at the completion of each treatment.

______

98. Explain why the client must sign a release statement.

______

99. Explain what to do if hair is lightened.

______

SOFT CURL PERMANENT

100. Describe a soft curl permanent. ______

101. Name two types of hair that cannot be treated with products containing ammonium thioglycolate.

1. ______

2. ______

IMPLEMENTS AND MATERIALS

102. List the implements and materials required to give a soft curl permanent.

1. ______________________ 2. ______________________
3. ______________________ 4. ______________________
5. ______________________ 6. ______________________
7. ______________________ 8. ______________________
9. ______________________ 10. ______________________
11. ______________________ 12. ______________________
13. ______________________ 14. ______________________
15. ______________________ 16. ______________________
17. ______________________ 18. ______________________
19. ______________________ 20. ______________________

PROCEDURE

103. List two conditions that prohibit the use of permanent waving gel or cream.

1. ______________________

2. ______________________

104. Describe what is done to hair after the shampoo. ______________________

105. Identify what is used to remove tangles from hair. ______________________

106. Specify the number of sections into which to part the hair. ______________________

107. Describe where to apply protective cream, if required by manufacturer.

____

108. Identify an item worn by the cosmetologist. ______________________

109. Explain how to apply the thio gel or cream.

____

110. Identify the areas of the hair and on the head to begin application.

____

111. Explain how to comb the thio gel or cream.

____

____

112. Explain what to do when hair becomes supple and flexible.

____

113. Identify the number of sections into which to divide the hair. ______________________

114. Specify how much larger than the natural curl the rod must be.

____

115. Specify how many times the hair should encircle the rod. ______________________

116. Describe how to protect the client's skin after completing the wrap.

____

117. Explain how to apply thio gel, cream, or lotion to hair.

____

____

118. Identify what is used to cover the client's head. ______________________________

119. Specify the amount of time the client must sit under a pre-heated dryer.

____

____

120. Explain what to do if the test curl shows an undeveloped curl pattern.

____

____

121. Specify the water temperature required for rinsing the hair. ____________________

122. Explain what is done to each curl before applying neutralizer. __________________

123. Specify how many times to saturate each curl with neutralizer. ________________

124. Specify the amount of time to allow neutralizer to remain on curls.

____

____

125. Explain what to do with balance of neutralizer after rods are removed.

____

____

126. Specify the water temperature required for rinsing after neutralization. __________

127. List the three final steps in the procedure for a soft curl permanent.

1. ____

2. ____

3. ____

AFTERCARE

128. List three aftercare tips.

   1. ______
   2. ______
   3. ______

   ______

REVIEW OF SAFETY PRECAUTIONS

129. Explain what to do if hair is treated with sodium hydroxide.

   ______

130. Explain what to do if hair is colored with a metallic dye or compound henna.

   ______

131. Explain what to do with information regarding the analysis of hair and scalp.

   ______

132. Explain what to do if hair is bleached, tinted, or damaged.

   ______

   ______

133. Explain what to do if waving lotion or neutralizer gets into client's eye.

   ______

134. Explain how to ensure proper curl formation without damage. ______

135. Describe where to apply protective cream. ______

136. Describe how to complete the client's record card. ______

WORD REVIEW

ammonium thioglycolate
conditioner-filler
"no base" relaxer
protective base
pull test
strand test
chemical blowout
elasticity
overlap
protective cream
relaxer text
thio relaxer
chemical relaxer
finger test
porosity
processing cream
sodium hydroxide

MATCHING TEST

Insert the correct term or phrase in front of each definition.

ammonium thioglycolate
chemical blowout
chemical hair relaxing
finger test
predisposition test
pull test
relaxer test
retouch
soft curl permanent
strand test

1. ______________________________ a test that determines the degree of elasticity in the hair

2. ______________________________ application of the relaxer only to the new growth

3. ______________________________ a test that indicates how fast the natural curl is being removed

4. ______________________________ also called thio relaxer

5. ______________________________ method of permanently waving overly curly hair

6. ______________________________ test that indicates the reaction of the relaxer on the hair

7. ______________________________ the process of permanently rearranging the basic structure of overly curly hair into a straight form

8. ______________________________ a test that determines the degree of porosity in the hair

9. ______________________________ a combination of chemical hair straightening and hairstyling

RAPID REVIEW TEST

Place the correct word in the space provided in each sentence below.

abrasions
activator
after
allergic
before
cold
continues
dandruff
eyes
fingers
first
force
gloves
hot
infected
last
less
manufacturer
more
"no base"
oils
pick
pores
record
scalp
shampoo
stabilizer
temperature
three
top
two
underneath
water

1. Scratches on the scalp may become seriously ______________ when aggravated by the chemicals in the relaxer.
2. When using a sodium hydroxide relaxer, the hair is shampooed ____________ it is relaxed.
3. If permanent waving lotion or neutralizer accidentally gets into the client's eye, it must be flushed immediately with _____________
4. A lifting ____________ should be used on hair treated with a soft curl permanent.
5. Water that is too ____________ will not stop the processing action of the relaxer.
6. Cosmetologists must wear protective _____________.
7. Processing cream is applied ____________ to the scalp area and hair ends.
8. A _____________ of each chemical hair relaxing treatment helps to ensure consistent, satisfactory results.
9. The ____________ of the rinse water should be used to remove the relaxer.
10. A protective base is not necessary when using a _______________ relaxer.
11. Chemical hair relaxers are applied to hair with the back of a comb, a tint brush, or the _____________.
12. Conditioner or curl ________________ should be used daily on hair treated with a soft curl permanent.
13. Cosmetologists must always follow the directions of the _________________.
14. Throughout the relaxer service, the client's ____________ must be protected.
15. A conditioner applied to hair before setting helps to restore some of the natural ___________ to the scalp and hair.
16. When rinsing relaxer from the hair, the stream of water is directed from the ____________ to the hair ends.
17. To rearrange the curl pattern of the hair in a soft curl permanent, the rod must be at least ___________ times larger than the natural curl.
18. Chemical relaxers must never be given to clients whose scalps have ________________.
19. Relaxer is applied first to the ___________ side of the strand, and then _________________.
20. When using a thio relaxer, the hair is shampooed _____________ it is relaxed.
21. Water that is too ___________ may cause discomfort because of the sensitive condition of the scalp.
22. Permanent waving gel or cream must not be used on a client who has experienced an ______________ reaction to a previous perm.

23. Unless the relaxer is completely removed from the hair, its chemical action

____

24. Heat opens the __________ of the scalp and causes irritation or injury.

25. __________ processing time is required for hair ends and hair at the scalp.

MULTIPLE CHOICE TEST

Read each statement carefully, then write the letter representing the word or phrase that correctly completes the statement on the blank line to the right.

1. Hair that ruffles or feels bumpy
   a) has elasticity b) is porous
   c) is thick d) is curly ________

2. A hair strand that lies smoothly against the scalp during a strand test is
   a) overprocessed b) underprocessed
   c) sufficiently relaxed d) insufficiently relaxed ________

3. Caring for a soft curl permanent includes shampooing
   a) once a day b) once a week
   c) once a month d) as often as necessary ________

4. The temperature of the rinse water used to remove the relaxer should be
   a) hot b) warm
   c) tepid d) cool ________

5. Excessive stretching after a chemical relaxer can cause hair
   a) breakage b) discoloration
   c) reversion d) strength ________

6. During the curling portion of the soft curl permanent, the client must sit under a
   a) cool dryer b) warm dryer
   c) pre-heated dryer d) heat lamp ________

7. Products containing ammonium thioglycolate should not be used on hair that has been treated with
   a) temporary hair color b) semi-permanent hair color
   c) metallic dye d) aniline derivative hair color ________

8. Protective cream is applied to the client's hairline and around the
   a) mouth b) nose
   c) eyes d) ears ________

9. Smoothing the relaxer cream through the hair stretches the hair gently into
   a) a straight position b) a wavy position
   c) a curly position d) an overly curly position ________

10. During a soft curl permanent, the hair is relaxed enough when it becomes
    a) stiff and resistant b) stiff and flexible
    c) supple and flexible d) supple and resistant ________

11. A clearer view of the scalp is obtained by parting the hair into

a) 1/2" sections b) 1" sections
c) 1 1/2" sections d) 2" sections ________

12. Application of the neutralizing shampoo is repeated until the hair

a) smells good b) lathers well
c) looks coated d) looks shiny ________

13. A release statement protects the salon and the

a) client b) client's family
c) cosmetologist d) salon owner ________

14. Prior to applying a sodium hydroxide relaxer, a conditioner-filler is applied to hair when it is

a) dry b) towel-dried
c) damp d) dripping wet ________

15. During a soft curl permanent, each curl is neutralized

a) once b) twice
c) three times d) four times ________

16. Using chemical relaxers on lightened hair is

a) a good idea b) highly recommended
c) advisable d) not advisable ________

17. The processing time is speeded up near the scalp by body

a) odor b) oil
c) perspiration d) heat ________

18. In a soft curl permanent, the product used to curl the hair contains

a) hydrogen peroxide b) sodium hydroxide
c) ammonium thioglycolate d) aniline derivative ________

19. Relaxer or rinse water that gets into a client's eyes must be washed out and the client must be

a) left alone b) taken home
c) referred to a physician d) taken to another salon ________

20. During a retouch, the previously treated hair is protected by applying a

a) cream conditioner b) creamy shampoo
c) styling lotion d) styling mousse ________

21. To achieve a good curl formation in a soft curl permanent, hair should encircle the rod at least

a) 1 time b) 1 1/2 times
c) 2 times d) 2 1/2 times ________

22. Hair that reverts or "beads" away from the scalp during a strand test needs to

a) be cut b) be styled
c) be conditioned d) continue processing ________

23. Excessive heat after a chemical relaxer can cause hair to

a) break b) darken
c) revert d) become stronger ________

24. Products containing ammonium thioglycolate should not be used on hair that has been treated with

a) styling lotion b) hair conditioner
c) aniline derivative hair color d) sodium hydroxide products ________

25. A conditioner-filler is a product containing

a) protein b) carbohydrates
c) ammonia d) hydrogen peroxide ________

26. Neutralizing shampoo is manipulated by working with the fingers

a) on top of the hair b) underneath the hair
c) only at the scalp area d) only at the hair ends ________

27. The relaxer test is done on an area where the hair is

a) soft and resistant b) wiry and resistant
c) wiry and fragile d) soft and fragile ________

28. In a soft curl permanent, the product used to relax the hair contains

a) hydrogen peroxide b) sodium hydroxide
c) ammonium thioglycolate d) aniline derivative ________

29. The purpose of a protective base is to protect the client's

a) face b) eyes
c) neck d) scalp ________

30. A relaxer retouch can be done every

a) week b) 2 weeks
c) 4 weeks d) 6 to 8 weeks ________

31. Damaged hair may be returned to a more normal condition by a series of

a) permanent waves b) permanent hair colors
c) conditioning treatments d) chemical hair relaxers ________

32. After saturating hair with neutralizer, a comb is used to

a) keep hair neat b) keep hair straight
c) scratch the scalp d) pull and stretch the hair ________

33. If moisture or perspiration is present on the scalp before application of a sodium hydroxide relaxer, the client is placed under a

a) heat lamp b) hot dryer
c) warm dryer d) cool dryer ________

34. Hair that appears to stretch when pulled gently

a) has elasticity b) is porous
c) is thick d) is curly ________

35. The important consideration in the chemical blowout is to not over-

a) condition the hair b) relax the hair
c) curl the hair d) dry the hair ________

Also see *Milady's Standard Theory Workbook.*

Date ______________

Rating ______________

Text Pages 301–312

THERMAL HAIR STRAIGHTENING (HAIR PRESSING)

INTRODUCTION

1. Identify the length of time that a hair pressing generally lasts.

   ____

2. List two services for which hair pressing prepares the hair.

   1. ______________ 2. ______________

3. Explain how much curl is removed by the soft press. ______________

4. Describe how to accomplish the soft press.

   ____

5. Explain how much curl is removed by the medium press. ______________

6. Describe how to accomplish the medium press.

   ____

   ____

7. Explain how much curl is removed by the hard press. ______________

8. Describe how to accomplish the hard press.

   ____

ANALYSIS OF HAIR AND SCALP

9. List four conditions of hair and scalp that prohibit hair pressing.

   1. ______________ 2. ______________

   3. ______________ 4. ______________

10. Explain what to do if hair shows signs of neglect or abuse caused by faulty pressing, lightening, or tinting. ______________

11. Describe what can happen if dry, brittle hair is not corrected before pressing.

    ____

12. Identify the length that hair with good elasticity can be safely stretched.

____

13. Explain what happens if the hair's ability to absorb water (porosity) is normal.

____

14. List eight points to include in a hair and scalp analysis.
    1. _______________
    2. _______________
    3. _______________
    4. _______________
    5. _______________
    6. _______________
    7. _______________
    8. _______________

15. Explain why cosmetologists must be able to recognize individual differences in characteristics of hair and scalp.

____

____

____

16. Name two factors that determine variations in hair texture.
    1. _______________
    2. _______________

17. List two ways that help determine how to treat a client's hair.
    1. _______________
    2. _______________

18. Explain why coarse hair is difficult to press.

____

____

19. Name the type of hair that is least resistant to hair pressing. _______________

20. Explain how to avoid breaking fine hair.

____

21. Compare the number of layers found in coarse and medium hair and in fine hair.

____

____

22. Describe the feel of wiry, curly hair. _______________

23. Explain why wiry, curly hair is very resistant to hair pressing and requires more heat and pressure than other hair types.

____

SCALP CONDITION

24. Explain what to do if the scalp is normal.

____

25. Explain what to do if the scalp is tight and the hair coarse.

____

26. Describe the main difficulty with a flexible scalp.

____

____

RECORD CARD

27. List four hair treatments to question client about.

1. ______________________ 2. ______________________

3. ______________________ 4. ______________________

CONDITIONING TREATMENTS

28. List three requirements of an effective conditioning treatment.

1. ____

2. ____

3. ____

29. List three services that help to render a tight scalp more flexible.

1. ______________________ 2. ______________________

3. ______________________

PRESSING COMBS

30. Name two types of pressing combs.

1. ______________________ 2. ______________________

31. Describe the materials used to make pressing combs. ______________________

32. Explain why the handle is made of wood. ______________________

33. Compare the results obtained from a comb with more space between its teeth and a comb with less space between its teeth.

____

____

34. Compare the lengths of hair with which to use short and long pressing combs.

____

35. Name two ways to heat regular pressing combs.

1. ______________ 2. ______________

36. Describe the positions of the teeth and of the handle while the comb is being heated.

____

37. Describe what is used to test the pressing comb. ______________

38. Explain what to do if the paper becomes scorched.

____

39. Differentiate between the two forms of electric pressing combs.

____

____

40. Explain how to keep the pressing comb clean.

____

41. Explain how to remove the carbon from the pressing comb.

____

____

____

PRESSING OIL OR CREAM

42. List six beneficial effects of applying pressing oil or cream to hair.

1. ______________
2. ______________
3. ______________
4. ______________
5. ______________
6. ______________

HAIR SECTIONING

43. Specify the number of sections into which to divide the head. ______________

44. Explain what determines the size of the subsections. ______________

45. Name the size of subsections needed for each of the following hair types.

Medium texture, average density: ______________

Coarse hair, greater density: ______________

Fine or thin hair, sparse density: ______________

SOFT PRESSING PROCEDURE FOR NORMAL CURLY HAIR

46. List the equipment, implements, and materials required for a soft press.

    1. ______
    2. ______
    3. ______
    4. ______
    5. ______
    6. ______
    7. ______
    8. ______
    9. ______
    10. ______

47. List the eight steps required in the preparation for a soft press.

    1. ______
    2. ______
    3. ______
    4. ______
    5. ______
    6. ______
    7. ______
    8. ______

48. Specify the area of the head at which to begin the procedure.

    ______

49. Describe how to apply pressing oil over the small hair sections.

    ______

50. Explain how to determine the heat intensity of the comb before placing it on the hair.

    ______

51. Describe how to lift the end of a small hair section.

    ______

52. Describe where on the hair section to insert the teeth of the pressing comb.

    ______

53. Describe how to make the hair strand wrap itself partly around the comb.

    ______

54. Identify the part of the comb that actually does the pressing. ______

55. Describe how long to press the comb slowly through the hair strand.

____

____

56. Specify where to bring each completed hair section. ______________________

57. List the five steps required to complete the service.

1. ____
2. ____
3. ____
4. ____
5. ____

HARD PRESS

58. Explain when to recommend a hard press.

____

59. Describe how to do a hard press. ______________________

60. Give another name for a hard press. ______________________

TOUCH-UPS

61. Describe when touch-ups are necessary.

____

____

62. Identify the service that is omitted during the touch-up. ______________________

SAFETY PRECAUTIONS

63. List three injuries that are the immediate results of hair pressing.

1. ____
2. ____
3. ____

64. List two injuries that are not immediately evident.

1. ____

____

2. ____

____

65. Describe what to do in case of a scalp burn.

____

66. List four things cosmetologists must avoid to assure the client's safety.

1. ______________________________

2. ______________________________

3. ______________________________

4. ______________________________

RELEASE STATEMENT

67. Explain the purpose of a release statement.

____

____

____

REMINDER AND HINTS ON SOFT PRESSING

68. List the seven reminders and hints on soft pressing.

1. ______________________________

2. ______________________________

3. ______________________________

____

4. ______________________________

____

5. ______________________________

6. ______________________________

7. ______________________________

PRESSING FINE HAIR

69. Explain how to avoid breakage when pressing fine hair.

____

____

PRESSING SHORT, FINE HAIR

70. List two reasons why the pressing comb should not be too hot when pressing fine hair that is extra short.

1. ______________________ 2. ______________________

71. Describe what to do in the event of an accidental burn.

____

PRESSING COARSE HAIR

72. Explain why enough pressure must be applied when pressing coarse hair.

____

PRESSING TINTED, LIGHTENED, OR GRAY HAIR

73. Explain how to obtain good results when pressing tinted, lightened, or gray hair.

____

74. Describe what can happen if a hot comb or iron is used on tinted or lightened hair.

____

WORD REVIEW

carbon
gentian violet jelly
porosity
thermal hair straightening
elasticity
hair pressing
pressure
flexible
pomade
sheen

MATCHING TEST

Insert the correct term or phrase in front of each definition.

double comb press
soft press
hard press
medium press

1. ______________________ pressing the hair by applying the thermal pressing comb once on each side of the hair

2. ______________________ pressing the hair by applying the thermal pressing comb twice on each side of the hair

3. ______________________ another name for hard press

4. ______________________ pressing the hair by applying the thermal pressing comb once on each side of the hair using slightly more pressure

RAPID REVIEW TEST

Place the correct word in the space provided in each sentence below.

carbon	clean	coarse
conditioned	density	fine
flexible	greatest	hair
hard	hot	light
opposite	record	sheen
smallest	texture	tight
touch-up	wet	

1. Burnt hair strands cannot be ________________.

2. A ___________ press is recommended when the results of a soft press are not satisfactory.

3. ____________ hair with greater density requires small subsections.

4. ____________ is removed by rubbing the outside surface of the pressing comb with fine sandpaper, emery board, or steel wool.

5. Tinted, lightened, or gray hair should be pressed with ___________ pressure.

6. Hair that becomes curly again due to perspiration or dampness may require a ______________

7. Cosmetologists should keep a ____________ of all pressing treatments.

8. Coarse hair has the _____________ diameter; fine hair has the _____________ diameter.

9. The main difficulty with a ____________ scalp is that the cosmetologist might not apply enough pressure to press the hair satisfactorily.

10. After each use, the comb should be wiped clean of loose __________ grease, and dust.

11. Pressing oil or cream adds ___________ to pressed hair.

12. The size of subsections depends on the ____________ and ____________ of the hair.

13. Using a __________ comb or iron on tinted or lightened hair can cause discolaration or breakage.

14. As each hair section is pressed, it is placed on the _____________ side of the head.

15. Pressed hair returns to its normal curly appearance when it is __________.

MULTIPLE CHOICE TEST

Read each statement carefully, then write the letter representing the word or phrase that correctly completes the statement on the blank line to the right.

1. Hair may be pressed in the presence of
   a) a scalp abrasion b) damaged hair
   c) normal, healthy hair and scalp d) a contagious scalp condition ________

2. A soft press removes about
   a) 30–45% of the curl b) 50–60% of the curl
   c) 60–75% of the curl d) 100% of the curl ________

3. The type of hair that requires more heat and pressure than any other hair type is
   a) coarse hair b) medium hair
   c) fine hair d) wiry, curly hair ________

4. A client who is allergic to pressing oil may develop
   a) hair breakage b) burnt hair
   c) a skin rash d) burnt skin ________

5. Tinted, lightened, or gray hair should be pressed with a
   a) very hot comb b) hot comb
   c) moderately heated comb d) cold comb ________

6. Pressing combs are constructed of a good quality
   a) steel or brass b) aluminum or gold
   c) iron or bronze d) pewter or silver ________

7. A tight scalp can be rendered more flexible by systematic scalp massage and
   a) hair brushing b) hair coloring
   c) thermal hair straightening d) cold water rinses ________

8. Smoking or burning of hair during pressing is caused by
   a) not shampooing first b) conditioning first
   c) excessive application of pressing oil d) sparing application of pressing oil ________

9. Clients whose hair shows signs of neglect or abuse should receive a series of
   a) permanent waves b) chemical relaxers
   c) pressing treatments d) conditioning treatments ________

10. Wiry, curly hair may be recognized by its
    a) hard, glassy feel b) spongy feel
    c) silky feel d) soft feel ________

11. While the comb is being heated, its teeth should face
    a) downward b) upward
    c) toward the right side d) toward the left side ________

12. The size of subsections for thin or fine hair with sparse density should be
    a) very small b) small
    c) average d) large ________

13. An example of an injury that is not immediately evident is

a) burnt hair
b) burnt scalp
c) hair that breaks progressively
d) burns on ears and neck
____

14. Hair with good elasticity can be safely stretched about

a) one-fifth of its length
b) one-fourth of its length
c) one-third of its length
d) one-half of its length
____

15. Hair is prepared for a pressing treatment by applying

a) hair lacquer
b) chemical hair relaxer
c) permanent hair color
d) pressing oil or cream
____

16. The part of the comb that actually does the pressing is the

a) teeth
b) handle
c) back rod
d) front rod
____

17. A hair pressing treatment should not be given to a client with

a) a scalp injury
b) dandruff
c) coarse hair
d) fine hair
____

18. A document relieving the cosmetologist from responsibility for accidents or damages is the

a) record card
b) release statement
c) insurance policy
d) malpractice policy
____

19. The layer missing from fine hair is the

a) cuticle
b) cortex
c) medulla
d) follicular layer
____

20. The pressing comb is kept sterile by

a) frequent use
b) infrequent use
c) the intense heat
d) the pressing oil
____

21. Burns on the scalp can cause temporary or permanent hair

a) loss
b) growth
c) color
d) curl
____

22. The ability of hair to absorb water is called

a) texture
b) density
c) elasticity
d) porosity
____

23. Handles of pressing combs are usually made of

a) metal
b) stone
c) wood
d) a synthetic material
____

24. The best time to press hair is

a) before the shampoo
b) after the shampoo
c) after a permanent hair color
d) after a permanent wave
____

25. A hard press removes about

a) 30–45% of the curl
b) 50–60% of the curl
c) 60–75% of the curl
d) 100% of the curl
____

26. After heating the comb to the proper temperature, it is tested on a

a) brown towel
b) tan towel
c) piece of white paper
d) piece of dark paper
____

27. Smoking or burning of hair during pressing is prevented by

a) not shampooing first
b) conditioning first
c) drying completely after shampooing
d) towel-drying after shampooing
____

28. Scalp massage and direct high-frequency current help to stimulate

a) dandruff on the scalp
b) dryness of the scalp
c) gray hair
d) blood circulation
____

29. The type of hair that is least resistant to hair pressing is

a) coarse hair
b) medium hair
c) fine hair
d) wiry, curly hair
____

30. Pressing oil should be applied

a) sparingly
b) generously
c) liberally
d) excessively
____

31. Short pressing combs are meant to be used on

a) short hair
b) medium-length hair
c) long hair
d) very long hair
____

32. The process for a touch-up is the same as for the original pressing treatment, except that the

a) hair is tinted first
b) hair is shampooed first
c) hair is shampooed and blow-dried first
d) shampoo is omitted
____

33. Scalp burns should be treated immediately with

a) boric acid
b) hydrogen peroxide
c) 1% gentian violet jelly
d) 70% alcohol
____

34. A comb with more space between the teeth produces a

a) fine looking press
b) medium looking press
c) coarse looking press
d) smoother press
____

35. The size of subsections for medium-textured hair of average density should be

a) small
b) average
c) large
d) extra large
____

36. Progressive breaking and shortening of the hair is the result of

a) frequent haircuts
b) frequent conditioners
c) infrequent hair pressings
d) too frequent hair pressings
____

37. A medium press removes about

a) 30–45% of the curl
b) 50–60% of the curl
c) 60–75% of the curl
d) 100% of the curl
____

38. An example of an injury that is not immediately evident is

a) burnt hair
b) burnt scalp
c) burns on ears and neck
d) skin rash
____

39. The type of hair that requires less heat and pressure than other hair types is

a) coarse hair
b) medium hair
c) fine hair
d) wiry, curly hair
____

40. The metal portion of the comb will acquire a smooth and shiny appearance when immersed in a solution of

a) hydrogen peroxide
b) baking soda
c) disinfectant
d) rubbing alcohol
____

Date ____________

Rating ____________

Text Pages 313–330

THE ARTISTRY OF ARTIFICIAL HAIR

TAKING WIG MEASUREMENTS

1. Explain why an accurate measurement of the client's head is necessary.

   ____

2. Explain how to keep hair flat and tight to the scalp.

   ____

3. Name the implement used to measure the head. ____________________

4. List the six required wig measurements.
   1. ____________________
   2. ____________________
   3. ____________________
   4. ____________________
   5. ____________________
   6. ____________________

ORDERING THE WIG

5. Identify where to forward a copy of client's head measurements.

   ____

6. List four hair characteristics to include on the written record.

   1. ____________________ 2. ____________________

   3. ____________________ 4. ____________________

BLOCKING THE WIG

7. Describe the type of block that withstands continuous pinning and rough handling.

   ____

8. List the six sizes of canvas blocks.

1. ______________________ 2. ______________________
3. ______________________ 4. ______________________
5. ______________________ 6. ______________________

9. Describe two reasons for good blocking.

1. ____
2. ____

10. Explain what happens when a wig is stretched and pinned on a block that is too big (and the cap is wet). ____

11. Explain what happens when a wig is hung too loosely, on a block that is too small (and the cap, if made of cotton, is wet). ____

12. Name four places to pin the wig.

1. ______________________ 2. ______________________
3. ______________________ 4. ______________________

13. Describe what Styrofoam blocks are used for. ______________________

ADJUSTING THE WIG TO A LARGER SIZE

14. Explain how to stretch a wig that is too tight.

____

____

ADJUSTING THE WIG TO A SMALLER SIZE

15. Explain the purpose of making horizontal tucks.

____

16. Explain the purpose of making vertical tucks.

____

17. Explain why the fit of the cap should be checked after each tuck.

____

18. Explain what to do if the wig touches the ear.

____

19. Explain what to do if the wig rubs or touches the side of the ear.

____

20. Explain what to do if the wig is too long from forehead to nape.

____

21. Explain why tucks are sewn toward (never away from) the crown.

____

22. Describe what to do with the hair when tacking a tuck.

____

23. Identify where to sew the tucks when adjusting a ventilated or hand-tied wig.

____

24. Specify where to sew the tucks when adjusting a machine-made or wefted wig.

____

THE ELASTIC BAND

25. Describe the final step in the wig adjustment process.

____

26. Explain why some stylists prefer pinning the ends of the elastic band with a small safety pin.

____

27. Explain why the elastic band requires periodic adjustment or replacement.

____

____

28. Identify a method of securing the elastic band preferred by other stylists.

____

HUMAN HAIR WIGS

29. Identify how often a human hair wig needs dry-cleaning.

____

30. Explain how to protect the canvas block. ______________________________

31. Identify the purpose of brushing the hair. ______________________________

32. Describe how to mark the size of the wig on the block so that the wig doesn't stretch or shrink.

____

____

33. Specify the type of cleaner used to clean a human hair wig.

____

34. Identify which side of the wig is down when it is being dipped.

____

35. Describe the alternate method of cleaning.

____

____

36. Explain how to remove excess liquid. _______________

37. Explain what to do after removing excess liquid.

____

____

38. Specify when to set and style the wig. _______________

HAND-TIED WIGS

39. Explain why hand-tied wigs should be cleaned on a block.

____

40. List the 11 steps required to clean a hand-tied wig.

1. _______________
2. _______________
3. _______________
4. _______________
5. _______________
6. _______________
7. _______________
8. _______________
9. _______________
10. _______________
11. _______________

SYNTHETIC WIGS

41. Explain why synthetic wigs and hairpieces need not be cleaned as often as human hair wigs. _______________

42. Specify how often synthetic wigs and hairpieces need to be cleaned.

____

43. Describe what happens when hot water is used to clean synthetic wigs.

____

44. Identify what not to do to a synthetic wig when it is wet. _______________

45. List the 11 steps necessary to clean a synthetic wig.

1. ______
2. ______
3. ______
4. ______
______
5. ______
______
6. ______
7. ______
8. ______
9. ______
10. ______
11. ______

46. Explain how to reduce drying time. ______

47. Explain why most synthetic wigs require no further styling. ______

CONDITIONING THE WIG

48. Identify something lacking in a wig that is present in natural human hair.

______

49. List two reasons for using a conditioning treatment after each wig cleaning.

1. ______
2. ______

50. Specify what implement is used to distribute the conditioner throughout the hair.

______

51. Specify the length of time to keep the conditioner in the hair.

______

SHAPING WIGS

52. Compare the amount of hair in a wig to the amount on a human head.

______

53. Describe the effect of not thinning and tapering a wig properly.

______

54. Name three areas to remove bulk on a wig.

1. ______________________ 2. ______________________

3. ______________________

55. Identify how close to the foundation the hair can be cut.

____

56. List two reasons for thinning hair close to the cap.

1. ____

2. ____

57. Explain why special care is necessary when cutting hand-knotted wigs.

____

58. Explain why special care is necessary when cutting wefted wigs.

____

59. Explain why caution is required when cutting wigs.

____

60. Explain why the client should wear the wig while the guideline is cut.

____

61. Explain why finishing the cut on a block is recommended.

____

62. Identify where to place the wig on the canvas block to cut it.

____

SHAPING SYNTHETIC HAIR WIGS

63. Explain why synthetic wigs should be cut when dry.

____

64. Explain why only scissors and thinning shears should be used on synthetic fibers or on a mixture of synthetic and human hair.

____

____

SETTING AND STYLING WIGS

65. List two ways in which setting wig hair differs from setting hair on the human head.

1. ______________________ 2. ______________________

66. Specify where wigs are always set and styled. ______________________

67. Explain how to keep the style close to the head.

____

68. Explain how to hold both rollers and curls more securely.

___

69. Explain why synthetic wigs may be teased or back-combed only at the base of their fibers.

___

70. Explain what happens when there is damage to the hair shaft.

___

71. Explain how the styles of synthetic wigs are determined.

___

PUTTING ON AND TAKING OFF A WIG

72. Describe how to remove a wig.

___

___

COLOR RINSES

73. Specify how often to reapply color rinses to human hair wigs.

___

74. Explain the limitations of color rinses. ___

75. Explain what to do if the client desires a lighter shade.

___

76. List the six steps required to apply a color rinse to a human hair wig.

1. ___
2. ___
3. ___
4. ___

___

5. ___
6. ___

77. Identify another type of hair color that can be used on human hair wigs.

____

78. Identify a type of hair color that should never be used on a wig or hairpiece.

____

SEMI-PERMANENT TINTS

79. Describe three characteristics of semi-permanent tints.

1. ____
2. ____
3. ____

80. List the nine steps necessary to apply semi-permanent tints to machine-made wigs.

1. ____
____
2. ____
3. ____
4. ____
5. ____
____
6. ____
7. ____
8. ____
9. ____

PERMANENT TINTS

81. Explain why applying a permanent tint to 100% human hair wigs and hairpieces is very risky. ____

82. Describe the result from applying permanent tint to 100% human hair wigs and hairpieces.

____

HAIRPIECES AND EXTENSIONS

83. Define switches. ______________________________

84. Explain how switches are worn.

____

85. Define wiglets.

____

86. Explain how wiglets are primarily used.

____

____

87. Define the bandeau-type hairpiece. ______________________________

88. Describe how the bandeau-type hairpiece is usually worn.

____

89. Define fall.

____

____

90. Name the three types of falls and specify their lengths.

1. ______________________ 2. ______________________
3. ______________________

91. Define demi-fall or demi-wig.

____

92. Specify the range in length of the demi-fall or demi-wig. ______________________

93. Define cascade.

____

94. Describe how cascades can be styled.

____

95. Define braid. ______________________________

96. Explain why some braids are prepared with a thin wire inside.

____

97. Define chignon.

____

____

98. Specify how a chignon works best. ______________________________

____

99. Define crown curls. ______________________________

100. Define frosting curls.

____

____

101. Identify, by labeling, the illustrated hairpieces.

______________ ______________ ______________

______________ ______________

SAFETY PRECAUTIONS

102. List the 11 safety precautions associated with artificial hair.

1. ______________________________
2. ______________________________
3. ______________________________
____
4. ______________________________
____
5. ______________________________
____
6. ______________________________
____
7. ______________________________
8. ______________________________
9. ______________________________
10. ______________________________
11. ______________________________

WORD REVIEW

artificial
braid
cascade
crown curls
extensions
frosting curls
machine-made
Styrofoam block
T-pins
wefted

bandeau
canvas block
chignon
demi-fall
fall
hand-tied
nonabsorbent
switch
tucks
wiglet

block
cap
circumference
elastic band
foundation
human hair
nonflammable
synthetic
ventilated

MATCHING TEST

Insert the correct term or phrase in front of each definition

bandeau	braid	cascade
chignon	crown curls	demi-fall
fall	frosting curls	switch
wiglet		

1. ____________________ a section of hair, machine wefted on a round base, available in various lengths
2. ____________________ a knot or coil of hair that works best when worn in combination with another hairpiece
3. ____________________ a hairpiece on an oblong base that offers an endless variety of styling possibilities
4. ____________________ a long weft of hair mounted with a loop at the end
5. ____________________ a hairpiece sewn to a headband
6. ____________________ a switch whose strands are woven, interlaced, or entwined
7. ____________________ a group of light curls worn on top of the head
8. ____________________ a large-base hairpiece designed to fit the shape of the head
9. ____________________ a hairpiece with a flat base used in special areas of the head

RAPID REVIEW TEST

Place the correct word in the space provided in each sentence below.

artificial	block	bulky
cleaners	damage	elastic band
foundation	index finger	measurements
natural	oils	plastic
problems	soft	thumb
tight	tucks	wefts
wet		

1. Accurate ____________________ of the client's head must be taken to assure a comfortable and secure fit.

2. Wigs that are not thinned and tapered properly will look ____________ and ______________.

3. When removing a wig from a client, only the ____________ is placed under the cap at the nape.

4. Too much tucking can cause the wig to ride up and create new fitting ______________.

5. Synthetic fibers can stretch out of shape if pulled when __________.

6. The block is protected by covering with a ____________ cover.

7. A wig differs from natural human hair in that it does not have its own supply of natural __________ for self-lubrication.

8. Cutting and thinning may be done as close to the wig ________________ as possible.

9. The final step in adjusting a wig is to adjust the _________________ at the back of the wig.

10. ______________ to the hair shaft causes a wig to become frizzy and fuzzy looking.

11. Because of the delicate structure of hand-tied wigs, they should be cleaned on a ___________.

12. When cutting a wig, it's important to avoid cutting the ____________ or sewing threads.

13. ____________ must be sewn toward the crown to prevent hair from standing away from the wig.

14. Wig ______________ usually dry the hair excessively.

15. A wig that is too ___________ may be adjusted by wetting its foundation with hot water, then stretching and pinning it carefully onto a larger size block.

MULTIPLE CHOICE TEST

Read each statement carefully, then write the letter representing the word or phrase that correctly completes the statement on the blank line to the right.

1. Brushing a synthetic wig when it is wet can

   a) discolor the strands b) dissolve the strands
   c) break the strands d) remove the curl ________

2. Compared to a human head, a wig holds about

a) half as much hair
b) the same amount of hair
c) twice as much hair
d) three times as much hair ________

3. Permanent damage to a synthetic wig can be caused by cutting it with a

a) pair of scissors
b) razor
c) thinning shear
d) texturizing shear ________

4. Hair in a wig or hairpiece cannot

a) be cleaned
b) be cut
c) be styled
d) grow back ________

5. When working with cleaning fluid, cosmetologists should wear

a) protective gloves
b) an apron
c) a protective head covering
d) a lab coat ________

6. Human hair wigs should be cleaned about every

a) week
b) 2 to 4 weeks
c) 6 to 8 weeks
d) 10 to 12 weeks ________

7. A written record of the client's head measurements is kept in the salon and a copy forwarded to the

a) client
b) client's spouse
c) wig dealer or manufacturer
d) state cosmetology board ________

8. A human hair wig may be treated safely with a

a) color rinse
b) hair lightener
c) permanent tint
d) permanent wave ________

9. To reduce the drying time of a synthetic wig or hairpiece, it may be placed under a

a) heat lamp
b) hot dryer
c) warm dryer
d) cool dryer ________

10. Rollers and pin curls are held securely with

a) T-pins
b) clippies
c) bobby pins
d) hair pins ________

11. Human hair wigs are cleaned with

a) an acid-balanced shampoo
b) an alkaline shampoo
c) a non-flammable liquid cleanser
d) a dandruff shampoo ________

12. Teasing or back-combing should be done only at the

a) crown area
b) top area
c) hair ends
d) base of the fibers ________

13. Synthetic wigs should be cut when they are

a) dripping wet
b) damp
c) slightly damp
d) dry ________

14. Applying a permanent tint to 100% human hair wigs and hairpieces results in

a) even color
b) uneven color
c) hair of good condition
d) shiny hair
____

15. Setting wig hair resembles setting hair on the human head except for coverage at the

a) top
b) crown
c) hairline
d) back
____

16. Synthetic wigs and hairpieces should be cleaned about every

a) 3 months
b) 6 months
c) 9 months
d) 12 months
____

17. Blocks used for all professional wig services are made of

a) wood
b) canvas
c) plastic
d) Styrofoam
____

18. To keep the style close to the head, the hair is set with

a) pin curls
b) small rollers
c) large rollers
d) a curling iron
____

19. Excess heat can cause synthetic hair strands to stick together and

a) burn up
b) dissolve
c) become curlier
d) lose their curl
____

20. A wet wig should always be mounted on a block of the

a) same head size
b) next samller head size
c) next larger head size
d) much larger head size
____

21. The best time to cut the guideline is while it is on

a) the block
b) the client's head
c) the cosmetologist's head
d) anyone's head
____

22. A part of the wig that requires periodic adjustment or replacement is the

a) elastic band
b) crown area
c) front hairline
d) sides
____

23. Synthetic wigs are pre-cut into definite styles by their

a) owners
b) owners' stylists
c) owners' friends
d) manufacturers
____

24. To prevent hair from becoming dry and brittle, wigs should be

a) tinted
b) lightened
c) reconditioned
d) permanently waved
____

Also see *Milady's Standard Theory Workbook.*

Date ______________

Rating ______________

Text Pages 331–362

MANICURING AND PEDICURING

PREPARATION OF THE MANICURING TABLE

1. Specify what rules must be followed to give a professional manicure.

____

2. Explain how to put the client in a more receptive mood for professional advice and suggestions.

____

____

PROCEDURE

3. List the seven steps necessary to set up the manicure table.

   1. ____
   2. ____
   3. ____
   4. ____
      ____
   5. ____
      ____
   6. ____
   7. ____

4. Identify what the manicuring table drawer should not be used for. ______________

5. Identify the manicuring implements and materials in the illustration.

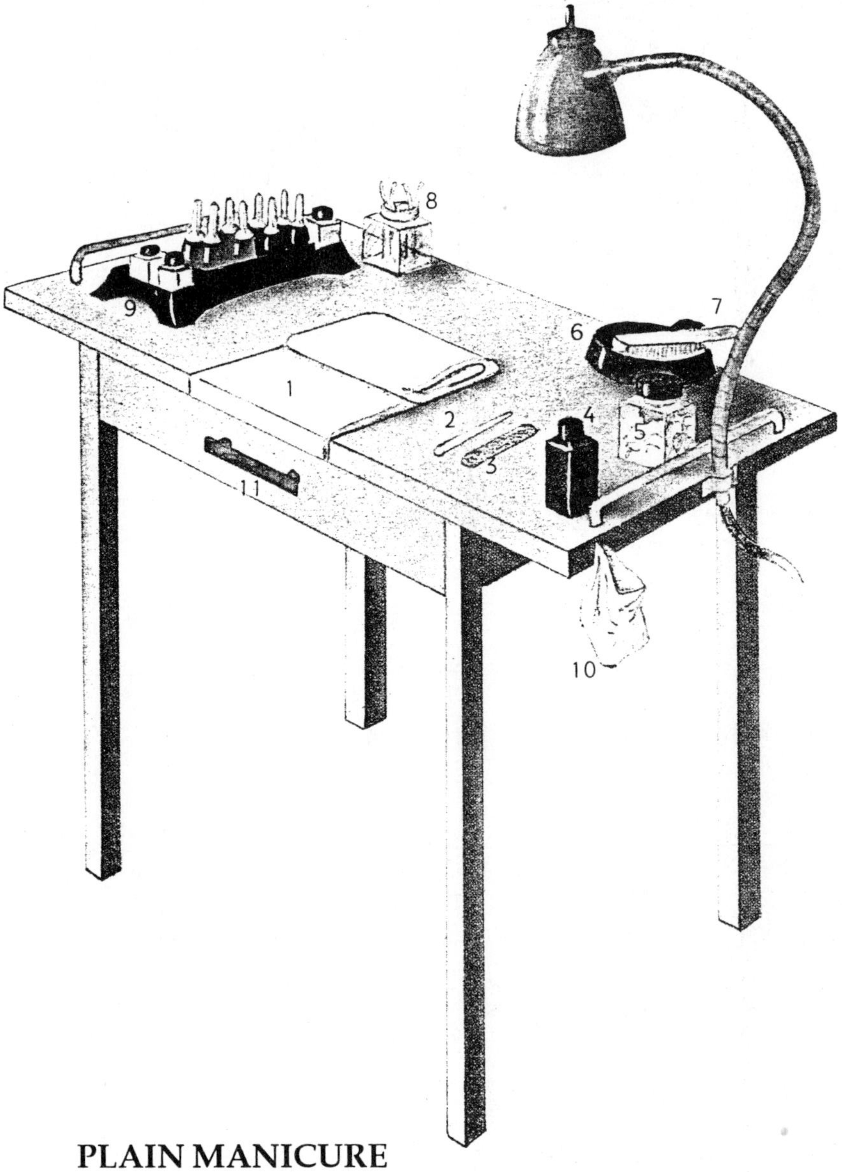

1. ______________________
2. ______________________
3. ______________________
4. ______________________
5. ______________________
6. ______________________
7. ______________________
8. ______________________
9. ______________________
10. ______________________
11. ______________________

PLAIN MANICURE

6. List the five steps required in the preparation for a manicure.
    1. ______________________
    2. ______________________
    3. ______________________
    4. ______________________
    5. ______________________

7. Explain what to do if the client's skin is cut during the manicure.

____

PROCEDURE

8. Specify where to begin removing old polish. ______________________

9. Describe how to soften the old polish.

____

____

10. Explain an alternate method of removing old polish.

____

____

____

11. Specify where to begin filing nails.

____

12. Describe how to hold the client's finger.

____

13. Explain how to confine the filing to the underside of the free edge.

____

14. Explain how to avoid splitting nails while filing.

____

15. List two reasons for avoiding filing deep into the corners of the nails.

1. ____________________ 2. ____________________

16. Explain when to immerse the left hand into the finger bowl.

____

17. Specify the reason for immersing the left hand into the finger bowl.

____

18. Describe how to treat the cuticles while drying the fingertips.

____

19. Describe how to make an applicator to use for applying cuticle remover.

____

20. Identify which end of the cuticle pusher is used to loosen cuticle. ____________

21. Describe how the cuticle must be kept while loosening it. ____________

22. Specify in what position the cuticle pusher is used while removing dead cuticle. ________

23. Explain how to avoid injury to the tissue at the root of the nail.

____

24. Specify what is used to clean under the free edge.

____

25. Specify in which direction to clean under the free edge. ____________

26. Name three things to remove with cuticle nippers.

    1. ______________________ 2. ______________________

    3. ______________________

27. Describe how the cuticle should be removed. ______________________

28. Specify when to immerse fingers of right hand into finger bowl.

    ____

29. Describe how to bleach under the free edge.

    ____

    ____

30. Explain how to apply nail whitener under free edge of nails.

    ____

    ____

31. Describe how to massage cuticle oil or cream into cuticles.

    ____

32. Explain how to manicure the nails and cuticles of the right hand.

    ____

33. Describe how to clean the nails of both hands.

    ____

34. Explain what to do after cleaning the nails of both hands.

    ____

NAIL PROBLEMS

35. Explain what causes a fringe of loose skin around the nail.

    ____

36. Describe how to prevent the occurrence of loose skin.

    ____

37. List two ways to soften callus growth at the fingertips.

    1. ______________________

    2. ______________________

38. Explain what is helpful to start the process of removing the callus.

    ____

39. List two ways to remove stains from fingernails.

    1. ______________________

    2. ______________________

COMPLETION

40. Describe what to use to give nails a smooth beveled edge.

____

41. Describe what may be given as an added service.

____

42. Identify the strokes used to apply base coat. _______________

43. Describe how long the base coat should be allowed to dry.

____

44. Explain how to wipe off excess liquid polish from the camel's hair brush.

____

45. Describe how to apply polish to nails.

____

46. Explain how to thin polish that is too thick. _______________

47. Describe how to remove excess polish from cuticles and nail edges.

____

____

48. Explain how to obtain added support and protection with top or seal coat.

____

49. Describe an additional service that may be given.

____

FINAL CLEANUP

50. List the six steps necessary for the final cleanup.

1. _______________
2. _______________
   ____
3. _______________
4 _______________
5. _______________
6. _______________

SAFETY RULES IN MANICURING

51. Explain how to prevent accidents and injury to the client or nail technician.

____

52. List the 13 safety rules associated with manicuring.

1. ______________________
2. ______________________
3. ______________________
4. ______________________
5. ______________________
6. ______________________
7. ______________________
8. ______________________
9. ______________________
10. ______________________

____

11. ______________________
12. ______________________
13. ______________________

INDIVIDUAL NAIL STYLING

53. Explain why the shape of the nail should conform to that of the fingertip.

____

54. List the four types of nail shapes.

1. ______________ 2. ______________

3. ______________ 4. ______________

55. Identify the nail shape considered to be the ideal. ___________

56. Specify the hand best suited for the slender tapering nail. ______________

57. Describe how to enhance the slender appearance of the hand.

____

58. Describe how far the square or rectangular nail should extend.

____

59. Describe how to shape the clubbed nail.

____

HAND MASSAGE

60. Explain the reason for including a hand massage with each manicure.

____

PROCEDURE

61. Specify where to spread the hand lotion. ______________________

62. Describe how to limber the client's wrist.

____

63. Describe how to limber the top of the hand and finger joints.

____

64. Describe how to completely relax the client's hand.

____

____

65. Specify where to grasp each finger before rotating in large circles. ______________

66. Describe three movements involved in manipulating the wrist.
    1. ______________________
    2. ______________________
    3. ______________________

67. Describe how to finish the hand massage.

____

____

HAND AND ARM MASSAGE

68. Identify how far to extend the hand and arm massage.

____

PROCEDURE

69. Describe (briefly) the seven steps required in the procedure for the hand and arm massage.
    1. ______________________
    2. ______________________

       ____
    3. ______________________
    4. ______________________
    5. ______________________
    6. ______________________

       ____
    7. ______________________

ELECTRIC MANICURE

70. Describe the device used to give an electric manicure.

____

____

71. Explain what cosmetologists must do before using an electric manicure machine.

____

OIL MANICURE

72. List three conditions that are improved by the oil manicure.
    1. ____________________ 2. ____________________
    3. ____________________

73. Describe how an oil manicure improves the hands.

____

74. Identify the type of oil used in the oil manicure.

____

75. Describe when to have client place fingers in the heated oil.

____

____

76. List three products that are not needed during the oil manicure.
    1. ____________________ 2. ____________________
    3. ____________________

77. Explain how to remove oil from the hands. ____________________

78. Explain how to remove oil from nails before applying base coat.

____

MEN'S MANICURE

79. Identify two nail shapes preferred by men.
    1. ____________________ 2. ____________________

80. Identify the type of polish used with a buffer. ____________________

81. Describe the strokes needed to buff the nails.

____

82. Explain how to prevent a heating or burning sensation on the nails.

____

83. List three benefits of buffing.

1. ______________________________

2. ______________________________

3. ______________________________

84. Specify when buffing is not required. ______________________________

BOOTH MANICURE

85. Describe a booth manicure.

____

ADVANCED NAIL TECHNIQUES

86. List six types of advanced nail techniques.

1. ______________ 2. ______________

3. ______________ 4. ______________

5. ______________ 6. ______________

87. List four reasons for using artificial nails.

1. ______________________________

2. ______________________________

3. ______________________________

4. ______________________________

NAIL WRAPPING

88. List two purposes of nail wrapping.

1. ______________ 2. ______________

89. Describe the difference between wrapping nails with silk and with linen.

____

____

90. Describe what must be done to help the mending material adhere to the nail.

____

91. Specify what is used to saturate the mending material. ______________

92. Indicate where to place the saturated mending material.

____

93. Describe how to smooth away the surface of the patch from the nail edge.

____

94. Indicate why a second patch may be necessary.

____

95. Describe what to do before applying the base coat and polish.

____

96. Explain how to roughen the nail surface before fortifying a nail.

____

97. Describe how the edges of the mending tissue should be torn.

____

98. Explain how to apply mending adhesive to mending tissue.

____

99. List two reasons for using the orangewood stick after placing the wrapping material over the nail.

1. _______________ 2. _______________

100. Describe what to do to the underside of the nail if using mending tissue.

____

____

101. Indicate where to apply one or two coats of adhesive.

____

102. Specify what to apply after the adhesive. _______________

103. Identify the type of product needed to loosen the nail wrap.

____

104. Identify which implement is used to gently remove the loosened wrap.

____

105. Specify where to place the client's fingertips after removing the wrap. _______________

106. Describe liquid nail wrap.

____

____

107. Explain how it is applied. _______________

108. Compare liquid nail wrap to nail hardener.

____

SCULPTURED NAILS

109. Give another name for sculptured nails. _______________

110. Explain when to use sculptured nails. _______________

111. List seven items needed to apply sculptured nails (in addition to those needed for manicuring).

1. ______________________ 2. ______________________
3. ______________________ 4. ______________________
5. ______________________ 6. ______________________
7. ______________________

112. Indicate the purpose of scrubbing the nails after giving the plain manicure.

____

113. Specify what is used to roughen the nail. ______________________

114. Identify what is used to dust the nail bed. ______________________

115. Specify whose directions to follow when applying nail primer to the nail surface.

____

116. Explain the purpose of pressing the nail form with the thumb and index finger.

____

117. Describe how the form should be under the free edge of the nail. __________

118. Explain how to form a smooth ball of acrylic.

____

____

____

119. Specify where to place the ball of acrylic. ______________________

120. Describe how to form the new acrylic nail tip.

____

121. Specify which area to avoid touching while the nail is being formed. __________

122. Describe how to make an extremely wet acrylic mixture. ______________

123. Specify when to remove the nail forms. ______________________

124. Explain the purpose of buffing the nails. ______________________

125. Identify the type of product required to remove sculptured nails.

____

126. Explain the correct way to remove sculptured nails.

____

____

127. Describe how not to remove sculptured nails.

____

128. Identify the type of product needed to remove polish from sculptured nails

____

129. Describe how to remove loose material from the nail bed.

____

130. Specify where to place the acrylic material. _______________

131. Describe how to blend the acrylic material into the existing sculptured nail.

____

132. Explain what to do if an acrylic nail is badly chipped or cracked.

____

133. List two reasons to use acrylic overlays.

1. _______________ 2. _______________

134. Identify where the nails are reinforced when using acrylic overlays.

____

135. List the six safety precautions associated with sculptured nails.

1. _______________
2. _______________
3. _______________
4. _______________
5. _______________
6. _______________

136. Explain what happens when sculptured nails lift, crack, or grow out and are not attended to immediately.

____

137. Explain what happens when the fungus spreads. _______________

138. Explain why manicurists should not treat the condition. _______________

139. Identify who should treat the condition. _______________

140. Indicate what a change in the color of the natural nail after sculptured nails have been applied usually means. _______________

141. Explain how the nail is sterilized before the acrylic is applied.

____

142. Explain how to help ensure that contamination does not occur.

____

PRESS-ON ARTIFICIAL NAILS

143. Identify two materials used to construct press-on nails.

 1. ______________________ 2. ______________________

144. List the three materials needed (in addition to those used in manicuring) to apply press-on nails.

 1. ______________________ 2. ______________________

 3. ______________________

145. Specify how far to give a manicure before applying press-on nails.

 ____

146. Explain how to roughen the client's nails. ______________________

147. Explain the purpose of trimming and filing the artificial nail at the cuticle end.

 ____

148. Describe how to flatten artificial nails. ______________________

149. Describe how to reshape artificial nails.

 ____

150. Identify where on client's nails — and where not — to apply adhesive for press-on nails.

 ____

151. Identify where on the artificial nails — and where not — to apply adhesive.

 ____

152. Specify the length of time needed to allow the adhesive to dry. ______________

153. Identify what the base of the artificial nail should touch when it is applied.

 ____

154. Specify the length of time needed to hold the press-on nail in place as it is applied.

 ____

155. Identify what should be wiped away from tips and around nails. ______________

156. Indicate the type of polish remover to use on nails made of plastic.

 ____

157. Explain what happens when polish remover containing acetone is used on plastic artificial nails. ____

158. Explain what kind of nails are not damaged when polish remover containing acetone is used on them. ____

159. Describe how to remove press-on artificial nails.

 ____

 ____

160. Explain why artificial nails should not be pulled or twisted off.

____

161. List two reasons to use adhesive solvent.

1. _______________
2. _______________

162. Describe how to care for press-on nails after removing from client's nails.

____

163. List five reminders and hints on press-on artificial nails.

1. _______________
2. _______________
3. _______________
4. _______________
5. _______________

DIPPED NAILS

164. Define dipped nails.

____

____

165. Indicate where glue is applied for dipped nails. _______________

166. Indicate what the nail is dipped into. _______________

167. Explain what is done to the nail when the acrylic dries.

____

168. Identify the type of product used to remove dipped nails.

____

NAIL TIPPING

169. Specify how far to give the manicure before applying tips.

____

170. Identify the part of the nail that requires roughening. _______________

171. Indicate what is filed to fit the shape of the free edge of the nail. _______________

172. Specify the amount of glue needed to apply to the free edge. _______________

173. Indicate the area of the nail on which to apply the tip. ______________________

174. Identify the area of the nail to buff.

____

175. Describe what to do with resulting dust. ______________________

176. Indicate where to apply nail glue. ______________________

177. Explain what happens when glue is placed over the nail dust.

____

178. Explain the purpose of filing the side of a nail tip. ______________________

179. Describe how to remove nail tips.

____

____

PEDICURE

180. Define pedicuring. ______________________

181. Name two benefits of good foot care.

1. ______________________ 2. ______________________

182. List three abnormal foot conditions that are best treated by a podiatrist.

1. ______________________ 2. ______________________

3. ______________________

183. Give another name for ringworm of the foot. ______________________

184. Define ringworm of the foot.

____

185. Specify who should treat clients with ringworm. ______________________

EQUIPMENT, IMPLEMENTS, AND MATERIALS

186. List the equipment, implements, and materials (in addition to those required for manicuring) needed for pedicuring.

1. ______________________ 2. ______________________

3. ______________________ 4. ______________________

5. ______________________ 6. ______________________

7. ______________________ 8. ______________________

9. ______________________ 10. ______________________

11. ______________________

PREPARATION

187. List the seven steps necessary in the preparation for pedicuring.

1. ______________________________
2. ______________________________
3. ______________________________
4. ______________________________
5. ______________________________
6. ______________________________
7. ______________________________

PROCEDURE

188. Describe how to file toenails.

____

189. Indicate how to avoid ingrown nails. ______________________________

190. Name two areas on which to apply cuticle solvent.

1. ______________ 2. ______________

191. Name the implement used to loosen the cuticle. ______________________________

192. Identify what is used to keep the cuticle moist. ______________________________

193. List two things to avoid when pushing back cuticle.

1. ______________ 2. ______________

194. Specify which part of the cuticle to nip. ______________________________

195. Indicate the type of product used to massage the toes. ______________________________

196. Describe the type of water needed to scrub both feet. ______________________________

FOOT MASSAGE

197. List the ten steps used in the procedure for a foot massage.

1. ______________________________
2. ______________________________
____
3. ______________________________
4. ______________________________
5. ______________________________
6. ______________________________
7. ______________________________
8. ______________________________
____
9. ______________________________
____
10. ______________________________

COMPLETION

198. List the six steps required to complete the pedicure.

1. ______________________________
2. ______________________________
3. ______________________________
4. ______________________________
5. ______________________________
6. ______________________________
____

LEG MASSAGE

199. Name two areas that should not be massaged when massaging from the ankle to the knee.

1. ____________________ 2. ____________________

200. Indicate where to apply pressure to the muscular tissue.

____

WORD REVIEW

acryllic

acrylic	adhesive	callus
cuticle	disinfectant	electric manicure
fill-in	fortify	hangnail
manicure	massage	nail wrap
non-acetone	oil manicure	overlay
pedicure	podiatrist	powdered alum
primer	pumice powder	ringworm
sculptured	shinbone	solvent

MATCHING TEST

Insert the correct term or phrase in front of each definition.

booth manicure	build-on nails	dipped nails
electric manicure	manicuring	nail tipping
nail wrapping	oil manicure	pedicuring
press-on nails		

1. ______________________ a procedure done to extend the natural nail artificially
2. ______________________ a manicure that is beneficial for ridged and brittle nails and for dry cuticles
3. ______________________ the care of the feet, toes, and toenails
4. ______________________ artificial nail tips that are sprayed with an adhesive and then applied with glue to the ends of the natural nails
5. ______________________ a manicure given with a portable device operated by a small motor
6. ______________________ another name for sculptured nails
7. ______________________ the care of the hands and nails
8. ______________________ reusable artificial nails that are convenient to apply
9. ______________________ a manicure that is not given at the manicuring table
10. ______________________ a process done to mend torn nails and to fortify weak nails

RAPID REVIEW TEST

Place the correct word in the space provided in each sentence below.

base	blood	callus
contamination	corners	cuticle
directions	dry	fingertip
flammable	flat	friction
fungus	hands	longer
manicure	massage	moist
nylon	pencil	plastic
polish	powder	pressure
segment	smooth	straight
stronger	tip	ventilated

1. For a more natural effect, the shape of the nail should conform to that of the _______________.
2. Styptic _____________ should not be used to stop bleeding.
3. A change in the color of the natural nail after the application of sculptured nails usually means that _____________ has become trapped under the acrylic.
4. While loosening cuticle, the cuticle is kept _____________.
5. Most artificial nail adhesives are _________________.
6. Buffing the nails increases circulation of ____________ to the fingertips.
7. Nails will look _____________ and be _______________ if permitted to grow out at sides.
8. A dab of hand lotion is applied to the client's hands before a _______________.
9. When trimming the cuticle, manicurists must be careful to remove it as a single _______________.
10. Polish removers containing acetone will damage artificial nails made of ______________.
11. Toenails should be filed _______________ across, with slightly rounded corners.
12. Manicurists who work with artificial nails should make sure the work area is well _________________.
13. When pushing cuticles, it's necessary to avoid too much _______________ at the base of the nail.
14. Touching the nails after application of a primer can cause ____________________.
15. Manicurists and pedicurists should not file too deeply into nail ______________.
16. Old polish is removed by pressing a cotton pledget moistened with remover over the nail for a few moments, then bringing it from ____________ of nail to __________.
17. When buffing nails, excessive _______________ should be avoided.
18. While loosening cuticle, the pusher is kept in a __________ position.

19. Polish removers containing acetone will not damage artificial nails made of _____________.

20. Manicurists must wash their _____________ before beginning the manicure.

21. A fringe of loose skin left around the nail after a manicure is caused by trimming the _____________ closer than necessary.

22. Hand massage keeps the hands flexible, well-groomed, and _____________.

23. Manufacturer's _______________ must be followed carefully when using manicuring products and equipment.

24. _____________ growth at the fingertips can be softened by removing the constant pressure that is causing it.

25. Liquid _____________ should be kept thin enough to flow freely.

MULTIPLE CHOICE TEST

Read each statement carefully, then write the letter representing the word or phrase that correctly completes the statement on the blank line to the right.

1. During the manicure, metal implements and orangewood sticks are kept
   a) on the manicure table b) in the drawer of the table
   c) in a jar sanitizer containing alcohol d) in the manicurist's pockets ________

2. The nail shape considered to be ideal is the
   a) oval b) square or rectangular
   c) slender tapering (pointed) d) clubbed (round) ________

3. Most manufacturers recommend that press-on nails not be worn longer than
   a) 2 weeks b) 4 weeks
   c) 6 weeks d) 8 weeks ________

4. Most primers have aseptic ingredients that
   a) color the nail b) roughen the nail
   c) soften the nail d) sterilize the nail ________

5. Clients with ringworm of the foot must be referred to a
   a) pedicurist b) manicurist
   c) cosmetologist d) physician ________

6. Stains on fingernails may be bleached with
   a) hair lightener b) prepared nail bleach
   c) household bleach d) laundry bleach ________

7. Manicures may not be given on nails when the surrounding skin is
   a) smooth b) rough
   c) well groomed d) inflamed or infected ________

8. Tops of nail polish bottles are cleaned with
   a) cuticle oil b) cuticle remover
   c) polish remover d) pumice powder ________

9. Removal of artificial nails should not include

a) applying polish remover
b) applying adhesive solvent
c) pulling or twisting off the nail
d) moisturizing skin afterward ________

10. To avoid splitting, nails are filed

a) from corner to center
b) from center to corner
c) from corner to corner
d) back and forth across nail ________

11. During leg massage, pressure is applied to the

a) muscular tissue
b) knee
c) shinbone
d) ankle ________

12. Before applying artificial nails, the surfaces of the natural nails are roughened with

a) adhesive
b) acrylic mixture
c) a cuticle pusher
d) an emery board ________

13. Two nail shapes preferred by most men are

a) oval and round
b) oval and pointed
c) round and square
d) square and pointed ________

14. A product similar to nail hardener that is thicker and contains more fiber is liquid

a) nail polish
b) nail wrap
c) base coat
d) top coat ________

15. To prevent a heating or burning sensation, the buffer is lifted from the nail after every

a) stroke
b) two strokes
c) three strokes
d) four strokes ________

16. The manicure table is wiped with disinfectant

a) only before the manicure
b) only after the manicure
c) before and after the manicure
d) only during the manicure ________

17. Cuts on the client's skin are treated with

a) 3% hydrogen peroxide
b) 6% hydrogen peroxide
c) nail polish remover
d) hand lotion ________

18. Immersing artificial nails in water for long periods may cause them to

a) harden
b) loosen
c) dissolve
d) strengthen ________

19. The implement used to clean under the free edge is

a) the pointed end of file
b) the pointed end of pusher
c) a cotton-tipped orangewood stick
d) any sharp, pointed implement ________

20. Fungus trapped under sculptured nails is

a) not contagious
b) very contagious
c) not of concern to the manicurist
d) not referred to a physician ________

21. The hand and arm massage extends to and includes the

a) elbow | b) upper arm
c) armpit | d) shoulder ________

22. The edges of mending tissue must be

a) blunt | b) straight
c) square | d) feathered ________

23. Before applying base coat, all traces of oil are removed with

a) cuticle solvent | b) hand lotion
c) polish remover | d) astringent ________

24. The type of nail that is applied over a nail form is the

a) nail wrap | b) sculptured nail
c) dipped nail | d) press-on nail ________

25. Men's manicures differ from women's in that men's are usually

a) more conservative | b) less conservative
c) more expensive | d) more involved ________

26. When holding or moving containers, the manicurist's hands should be

a) greasy | b) wet
c) damp | d) dry ________

27. The nail shape well suited for the thin, delicate hand is the

a) oval | b) square or rectangular
c) slender tapering (pointed) | d) clubbed (round) ________

28. Abnormal foot conditions, such as corns, calluses, and ingrown nails, are best treated by a

a) pedicurist | b) manicurist
c) cosmetologist | d) podiatrist ________

29. A nail wrap that gives a smooth, even appearance to the nail is done with

a) silk | b) linen
c) mending tissue | d) acrylic fiber ________

30. Waste materials are placed

a) on the floor | b) in a plastic bag
c) in the drawer of the manicure table | d) in the manicurist's pockets ________

Also see *Milady's Standard Theory Workbook*.

THE NAIL AND ITS DISORDERS

See *Milady's Standard Theory Workbook.*

THEORY OF MASSAGE

See *Milady's Standard Theory Workbook.*

Date ________________

Rating ________________

Text Pages 379–398

FACIALS

INTRODUCTION

1. Identify the service that is among the most enjoyable and relaxing to the salon client.

   ____

FACIAL TREATMENTS

2. List two reasons why cosmetologists must be able to recognize various skin ailments.
   1. ____
   2. ____

3. Describe preservative facials.

   ____

   ____

   ____

4. Describe corrective facials.

   ____

   ____

5. List ten benefits of facial treatments.
   1. ____
   2. ____
   3. ____
   4. ____
   5. ____
   6. ____
   7. ____
   8. ____
   9. ____
   10. ____

PREPARATION FOR FACIAL MASSAGE

6. Explain how to help the client relax. ______________________________

7. Describe how the cosmetologist must work to provide a quiet atmosphere.

____

8. Describe the conditions in the facial work area.

____

9. Identify what to use—and what not to use—to remove products from their containers.

____

10. Specify what to do with cold hands before touching the client's face. ______________

11. Describe how to avoid scratching the client's skin. ______________________________

EQUIPMENT, IMPLEMENTS, AND MATERIALS

12. List the items needed to give a facial.

1. ______________	2. ______________
3. ______________	4. ______________
5. ______________	6. ______________
7. ______________	8. ______________
9. ______________	10. ______________
11. ______________	12. ______________
13. ______________	14. ______________
15. ______________	16. ______________
17. ______________	18. ______________
19. ______________	20. ______________
21. ______________	22. ______________
23. ______________	24. ______________
25. ______________	

PROCEDURE

13. Explain the purpose of greeting the client and saying something complimentary.

____

14. Identify the items that the client must remove and that you must store in a safe place.

____

15. Explain the reason for placing a clean towel across the back of the facial chair.

____

16. Describe how to drape the client with the towel and coverlet.

____

____

____

17. Explain the reason for covering the client's head with a headband or towel before the facial. ____

18. Describe two types of head coverings on the market.

    1. ____
    2. ____

19. Describe how to fold the towel used for draping the head.

____

20. Describe how to place the towel over the headrest. _______________________

21. Identify where the back of the client's head should rest when the client is in a reclining position. ____

22. List three things to check after the towel is pinned in place.

    1. ____
    2. ____
    3. ____

23. Specify the position of the facial chair after lowering. _______________________

24. Identify what the cosmetologist must wash before starting the facial. ____________

25. List six determinations made by analyzing the skin.

    1. ____
    2. ____
    3. ____
    4. ____
    5. ____
    6. ____

26. List five purposes of analyzing the skin.

    1. ____
    2. ____
    3. ____
    4. ____
    5. ____

27. Identify the amount of cleansing cream or lotion to remove from the container.

____

28. Explain how to soften the cream or lotion. _______________________

29. Describe how to remove heavy eye or lip makeup.

____

30. Describe two ways to spread the cleansing product on various areas of the neck and face.

1. ____
2. ____

31. Describe the type of movements necessary to apply cleansing product to neck, chest, and back. ____

32. List four items that can be used to remove cleansing cream or lotion.

1. ____________________ 2. ____________________
3. ____________________ 4. ____________________

33. Identify the areas to begin and to finish removing the cleansing product.

____

34. Identify what can be done to eyebrows after the cleansing product is removed.

____

35. Name two ways to steam the face.

1. ____________________ 2. ____________________

36. Explain the purpose of steaming the face.

____

37. List two benefits of steaming the face.

1. ____
2. ____

38. Describe the type of massage cream to apply.

____

39. Describe how to apply massage cream. _______________________

40. Identify two areas to apply lubrication oil or cream (if needed).

1. ____________________ 2. ____________________

41. Describe what is used to cover the client's eyes during facial manipulations.

____

42. Specify when to expose the face to infrared light.

____

43. Describe what is used to cover the client's eyes during exposure to infrared light.

____

44. Identify the distance from the face to place the lamp. _______________

45. Identify the length of time to expose the face to infrared rays. _______________

46. Describe how to remove massage cream. _______________

47. Explain how to apply astringent or freshening lotion.

____

48. Identify the type of treatment mask to use on the client.

____

49. Identify the length of time to leave the mask on the face. _______________

50. Describe how to remove the mask. _______________

51. List two products that are applied after removing the mask.

1. _______________ 2. _______________

52. List five steps necessary to complete the facial.

1. _______________
2. _______________

____

3. _______________
4. _______________
5. _______________

FACIAL MANIPULATIONS

53. Identify what induces relaxation when giving facial manipulations.

____

54. Describe how to remove and how to replace hands once the manipulations have been started. _______________

55. Explain how to avoid damage to muscular tissues.

____

56. Label the following illustrations by filling in the correct names of the massage movements.

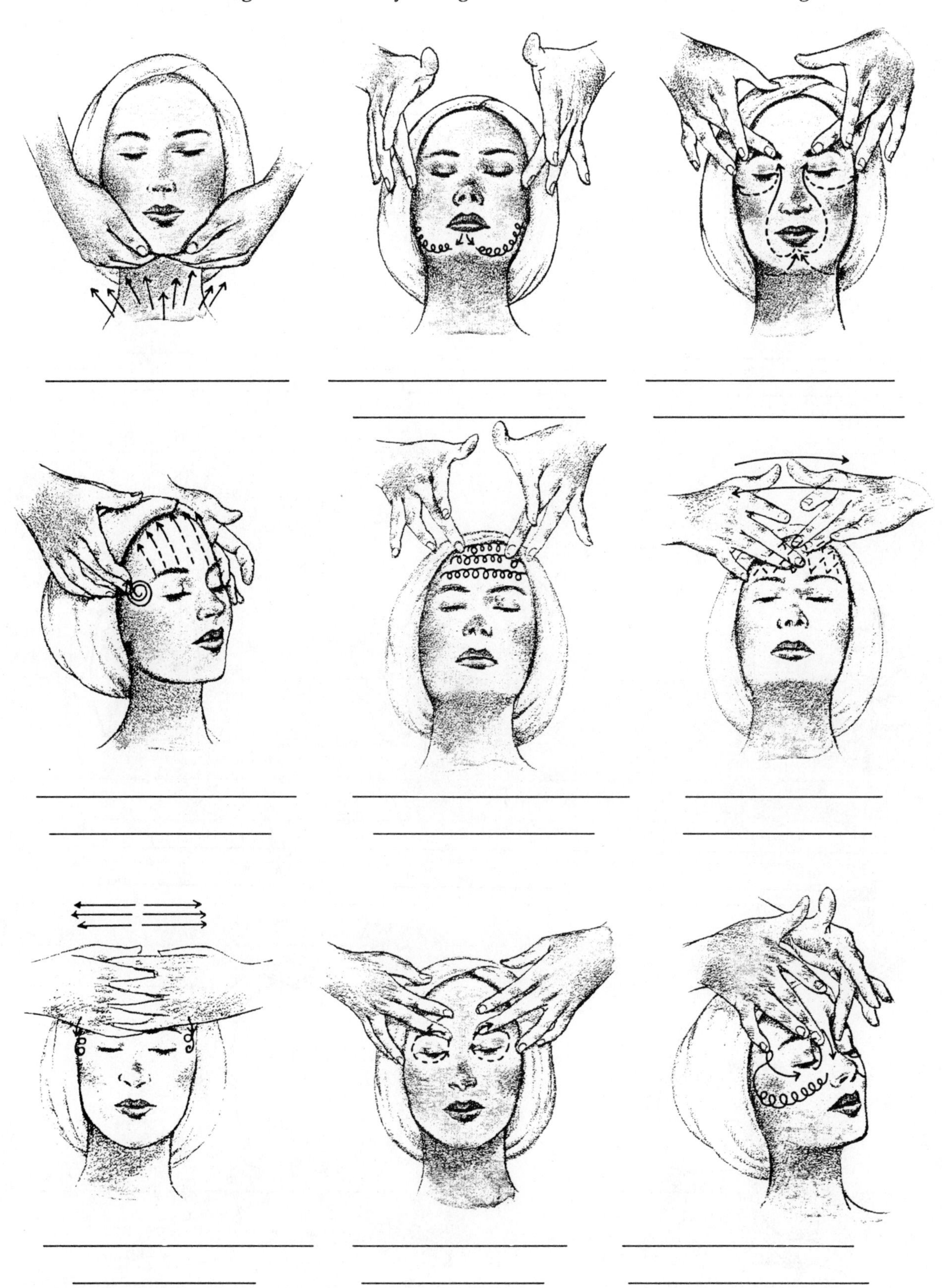

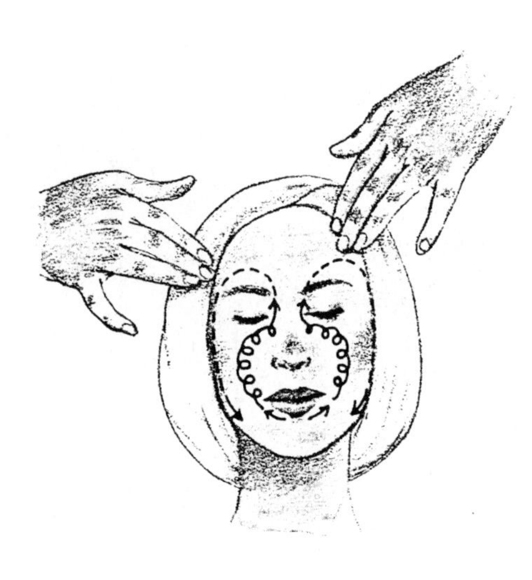

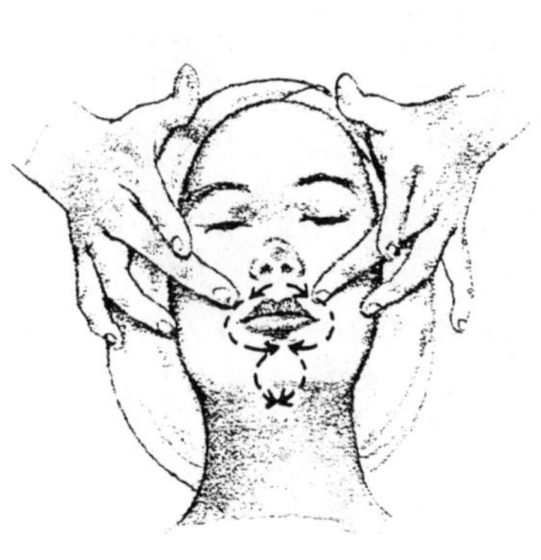

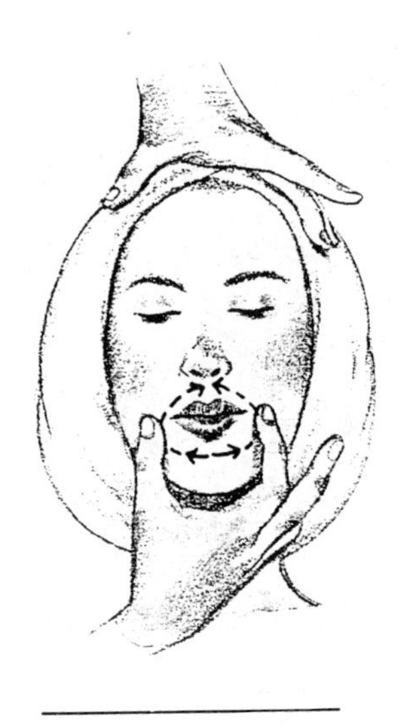

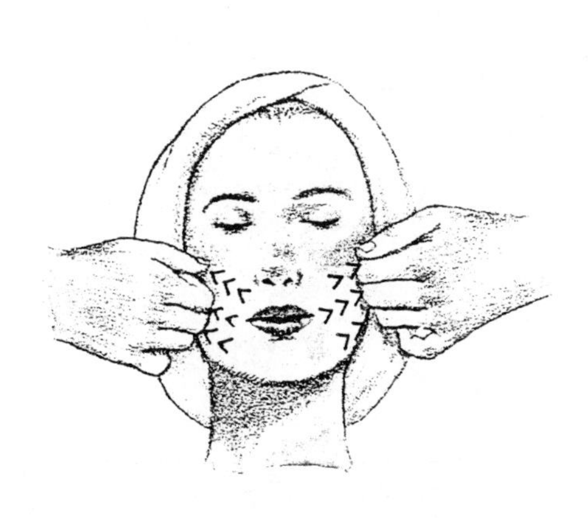

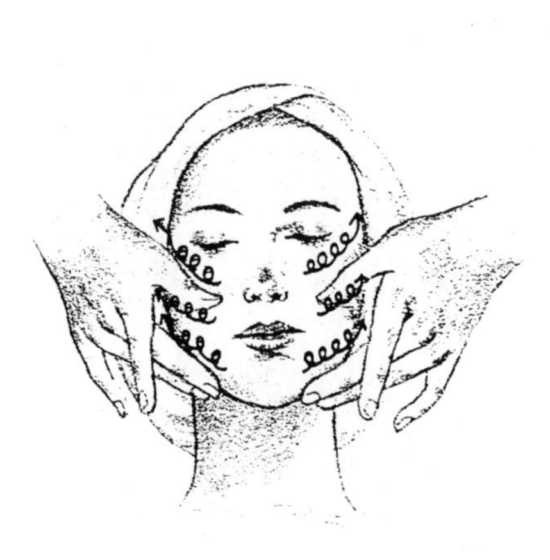

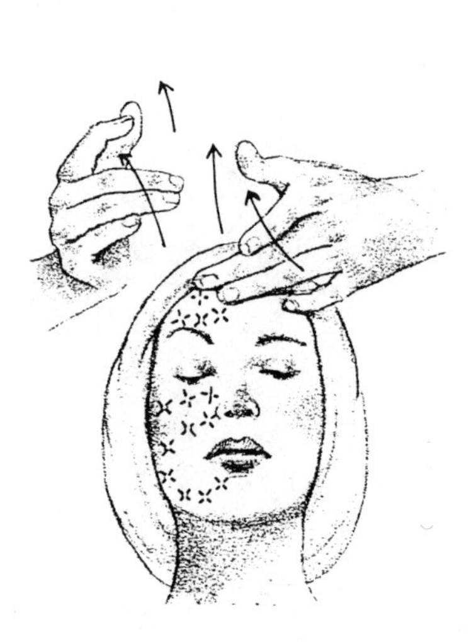

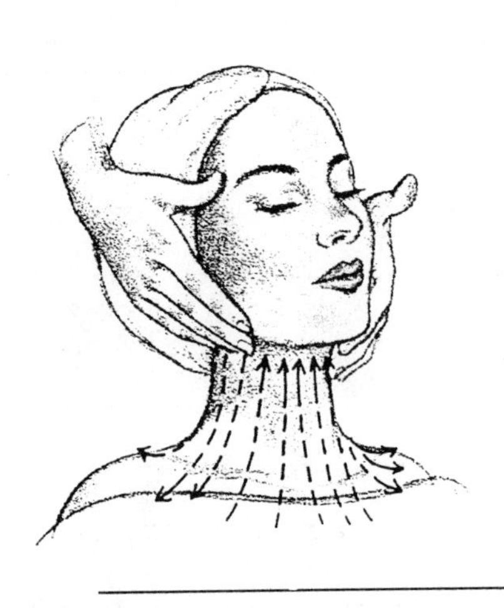

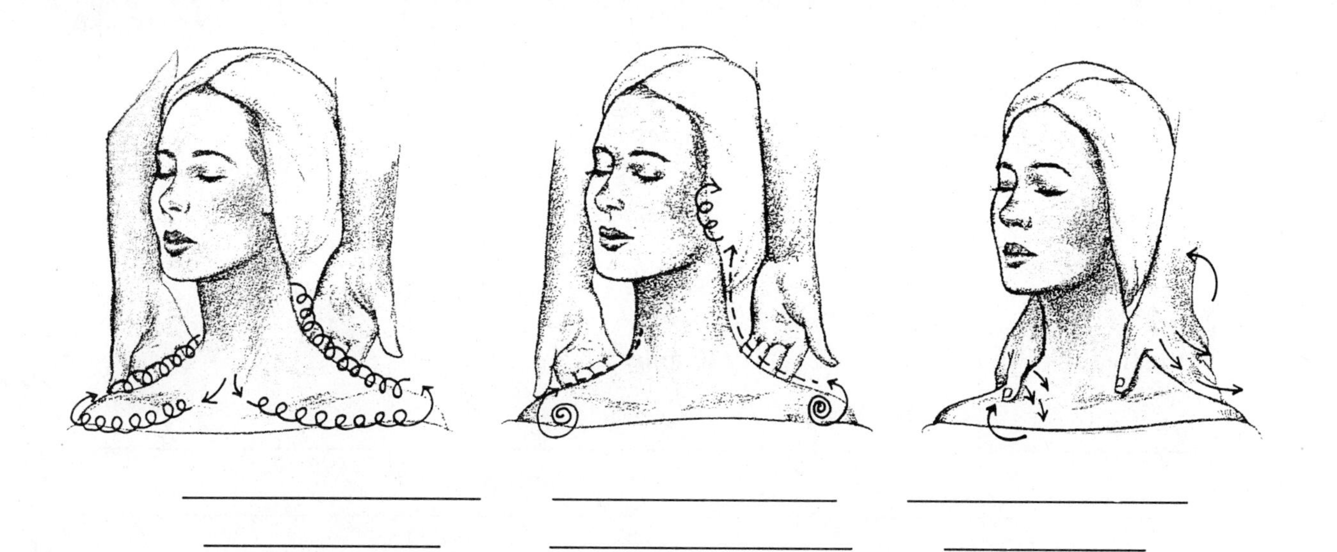

FACIAL FOR DRY SKIN

57. Identify the cause of dry skin.

58. Explain the purpose of the facial for dry skin.

59. Explain why the use of electrical current is recommended.

60. List the 14 steps in the procedure with infrared rays.
    1.
    2.
    3.
    4.
    5.
    6.
    7.
    8.
    9.
    10.
    11.
    12.
    13.
    14.

61. Identify the type of lotions to avoid using on dry skins.

____

62. List the five steps in the procedure with galvanic current.

1. ____

2. ____

3. ____

4. ____

5. ____

63. List the five steps in the procedure for dry skin facial with indirect high-frequency current.

1. ____

2. ____

____

3. ____

4. ____

5. ____

64. Describe the cause of an oily skin and/or blackheads (comedones).

____

65. Explain what the client can do to minimize oily skin and blackheads.

____

66. List the fifteen steps in the procedure for oily skin facial treatment.

1. ______________________________
2. ______________________________
____
3. ______________________________
____
4. ______________________________
____
5. ______________________________
____
6. ______________________________
7. ______________________________
8. ______________________________
9. ______________________________
10. ______________________________
11. ______________________________
12. ______________________________
____
13. ______________________________
14. ______________________________
15. ______________________________

WHITEHEADS (MILIA)

67. Describe what causes milia or whiteheads.

____

68. Explain why the condition usually occurs in fine-textured skin.

____

69. Identify the person under whose supervision this condition may be treated.

____

FACIAL FOR ACNE

70. Describe acne. ______________________________

71. Describe the role of the cosmetologist when the client is under medical care.

____
____

72. List four measures to which cosmetologists are limited when clients are under medical supervision for acne.

1. ______________________________
2. ______________________________
3. ______________________________
4. ______________________________

73. List the equipment, implements, and materials needed for an acne facial.

1. ____________________ 2. ____________________
3. ____________________ 4. ____________________
5. ____________________ 6. ____________________
7. ____________________

74. Explain why using rubber or latex gloves and disposable materials is advisable when treating acne skin. ______________________________

75. List the fourteen steps in the procedure for acne facial treatment.

1. ______________________________
2. ______________________________
3. ______________________________
4. ______________________________
5. ______________________________
____
6. ______________________________
7. ______________________________
8. ______________________________
9. ______________________________
____
____
10. ______________________________
____
11. ______________________________
12. ______________________________
13. ______________________________
14. ______________________________

DIET FOR ACNE

76. Name two factors believed to cause acne.

    1. ______________________ 2. ______________________

77. Name two things that can aggravate acne.

    1. ______________________ 2. ______________________

78. List three types of foods that tend to worsen an acne condition.

    1. ______________________ 2. ______________________
    3. ______________________

79. List three recommendations for an acne condition.

    1. ______________________
    2. ______________________
    3. ______________________

PACKS AND MASKS

80. Name two types of skin for which pack facials are recommended.

    1. ______________________ 2. ______________________

81. Describe how packs are usually applied. ______________________

82. Name the type of skin for which mask facials are recommended. ______________________

83. Describe how masks are usually applied.

    ____

84. Explain why gauze is often used with masks.

    ____

CUSTOM-DESIGNED MASKS

85. Explain when a custom-designed mask should not be used on a client.

    ____

86. Identify the length of time to leave a custom-designed mask on the face.

    ____

87. Name two benefits of a fresh strawberry mask.

    1. ______________________ 2. ______________________

88. Identify the type of skin for which a banana mask is recommended.

    ____

89. Name two qualities of tomatoes, apples, and cucumbers.

    1. ______________________ 2. ______________________

90. Name two effects of using egg white as a mask.

1. ______________________ 2. ______________________

91. Identify the effect of using yogurt or buttermilk as a mask.

____

92. List three effects of using honey as a mask.

1. ______________________ 2. ______________________

3. ______________________

93. Describe gauze. ______________________

____

94. Explain why gauze is used with masks.

____

95. Describe the type of ingredients that are applied over a layer of gauze.

____

96. Explain what the gauze does.

____

____

97. Specify the size of the piece of gauze needed.

____

98. Explain why spaces must be cut out for the eyes, nose, and mouth.

____

99. Describe how the mask ingredients are applied.

____

100. Explain how to remove the mask. ______________________

101. Identify the equipment, implements, and materials needed for a custom-designed mask.

____

102. Describe the eye pads used to protect the client's eyes.

____

103. Specify how far to progress with the facial before applying the mask.

____

104. Name four areas on which to avoid applying the mask.

1. ______________________ 2. ______________________

3. ______________________ 4. ______________________

105. Specify the length of time to allow the mask to remain on the skin.

____

106. Explain how to finish cleansing the face. ______________________

107. Identify two products to apply after the skin is cleansed.

1. ______________________________

2. ______________________________

108. Name the final step in the procedure for a custom-designed mask.

____

HOT OIL MASK FACIAL

109. List two types of skin for which a hot oil mask is recommended.

1. ______________ 2. ______________

110. Describe how to apply the mask to the skin.

____

111. List four items (in addition to those needed for a plain facial) needed to give a hot oil mask facial.

1. ______________ 2. ______________

3. ______________ 4. ______________

112. Specify how far to progress with the facial before applying the mask.

____

113. Identify the product used to moisten the eyepads.

____

114. Name what is used to moisten the gauze facial mask. ______________

115. Specify where to place the gauze mask. ______________

116. Specify the distance to place the infrared lamp from the client's face. ______________

117. Identify the length of time to allow client to rest under the lamp. ______________

118. List seven steps necessary to complete the procedure after the client has rested under the lamp.

1. ______________________________

2. ______________________________

3. ______________________________

4. ______________________________

5. ______________________________

6. ______________________________

7. ______________________________

REASONS A CLIENT MIGHT FIND FAULT WITH A FACIAL

119. A client might find fault with a facial treatment if you:

1. ______________________________
2. ______________________________
3. ______________________________
4. ______________________________
5. ______________________________

   ____
6. ______________________________
7. ______________________________
8. ______________________________
9. ______________________________
10. ______________________________
11. ______________________________
12. ______________________________
13. ______________________________
14. ______________________________

WORD REVIEW

acne
comedone
facial
high-frequency
infrared
massage
origin
spatula
tempo

allergic
corrective
galvanic
hot oil
manipulations
milia
pack
sebaceous

astringent
dermatologist
gauze
infectious
mask
moisturizer
preservative
sebum

MATCHING TEST

Insert the correct term or phrase in front of each definition.

acne
dry skin
milia
preservative facial

comedone
hot oil mask facial
oily skin

corrective facial
mask facial
pack facial

1. ______________________ a facial recommended for normal and oily skin, usually applied directly to the skin
2. ______________________ the technical term for blackhead
3. ______________________ a facial meant to maintain the health of the facial skin
4. ______________________ a skin condition caused by an insufficient flow of sebum from the sebaceous glands
5. ______________________ a facial recommended for dry skin, applied to the skin with the aid of gauze layers
6. ______________________ the technical term for whiteheads
7. ______________________ a facial meant to correct some facial skin conditions
8. ______________________ a disorder of the sebaceous glands requiring medical attention
9. ______________________ a facial recommended for dry, scaly skin, or skin that is inclined to wrinkle

RAPID REVIEW TEST

Insert the correct word in the space provided in each sentence below.

acne
astringent
diseases
hair
makeup
over

allergic
circulation
dry
insertion
nerves
physician

analysis
coarse
fine
jewelry
origin
under

1. To avoid damage to muscular tissues, massage movements are usually directed toward the ____________ of the muscle.
2. Skin ______________ helps to determine the areas of the face that need special attention.
3. When treating acne skin, cosmetologists need to work closely with the client's ______________.
4. Custom-designed masks are generally beneficial unless the client is ______________ to a particular substance.
5. The presence of acne or blackheads is more evident after ______________ is removed.
6. For more effective results, the use of electrical current is recommended for __________ skin.
7. Facial treatments are beneficial for relaxing the ____________.
8. Milia usually occurs in skin of __________ texture.
9. The client's __________ is protected by a headband, towel, or other head covering.
10. Eyepads are moistened with mild ______________.
11. ___________ may be worsened by eating foods high in fats, starches, and sugars.
12. Mask ingredients are applied __________ the gauze.
13. Cosmetologists do not treat skin ______________.
14. Before the facial begins, clients should remove their own ______________.
15. Facial treatments are beneficial for increasing ________________.

MULTIPLE CHOICE TEST

Read each statement carefully, then write the letter representing the word or phrase that correctly completes the statement on the blank line to the right.

1. Facial manipulations that induce relaxation are given with
   a) a fast tempo
   b) a jerky tempo
   c) an even tempo
   d) an uneven tempo ________
2. Sliced or crushed fruits or vegetables are held in place on the face with the aid of
   a) a plastic bag
   b) gauze
   c) a towel
   d) a coverlet ________
3. Cosmetologists must be able to recognize various skin ailments in order to advise clients to seek
   a) another cosmetologist
   b) another salon
   c) another manufacturer's cosmetics
   d) medical treatment ________

4. The facial work area must be

a) noisy
b) disorganized
c) sanitary
d) unsanitary
____

5. Products are removed from their containers with

a) combs
b) spatulas
c) the cosmetologist's fingers
d) the client's fingers
____

6. Some mask ingredients such as oatmeal can be mixed into a paste with

a) alcohol
b) astringent
c) hydrogen peroxide
d) milk
____

7. The first cream used in a plain facial is

a) massage cream
b) cleansing cream
c) emollient cream
d) moisturizing cream
____

8. Comedones are caused by a hardened mass formed in the ducts of the

a) sebaceous glands
b) sudoriferous glands
c) thyroid glands
d) lymph glands
____

9. Cosmetologists can help clients to relax by

a) speaking loudly
b) speaking quietly
c) talking constantly
d) playing loud music
____

10. Skin analysis helps to determine if lubricating oil or cream is needed around the

a) hairline
b) ears
c) eyes
d) nostrils
____

11. When treating dry skin, cosmetologists should avoid using products containing a large percentage of

a) banana
b) water
c) oil
d) alcohol
____

12. After draping the client and before analyzing the skin, cosmetologists must wash

a) their hands
b) the countertop
c) the facial chair
d) dirty towels
____

13. The amount of cleansing cream to apply to the face is about

a) a teaspoon
b) two teaspoons
c) a tablespoon
d) two tablespoons
____

14. Acne may be aggravated by

a) plenty of water
b) emotional stress
c) healthful personal habits
d) a well-balanced diet
____

15. A client is likely to find fault with a facial if the cosmetologist

a) is courteous
b) assists the client
c) has offensive breath or body odor
d) shows interest in the client
____

16. The pores of the skin are opened by applying

a) cold, moist towels
b) cold, dry towels
c) warm, moist towels
d) warm, dry towels
____

17. Cosmetologists should wear rubber or latex gloves when working on

a) sunburned skin b) acne skin
c) dry skin d) sensitive skin ________

18. Egg white is beneficial to

a) dry skin only b) normal skin only
c) oily skin only d) all skin types ________

19. The client's bare shoulders are prevented from coming in contact with the back of the facial chair by covering it with

a) a dirty towel b) a clean towel
c) the client's blouse d) the client's scarf ________

20. Cosmetologists can prevent scratching the client's skin by keeping nails

a) rough b) smooth
c) long d) polished ________

Date ____________

Rating ____________

Text Pages 399–432

FACIAL MAKEUP

PROCEDURE FOR APPLYING A PROFESSIONAL MAKEUP

1. Name five areas on which to apply dabs of cleanser.

   1. ____________ 2. ____________
   3. ____________ 4. ____________
   5. ____________

2. Describe the movements needed to spread the cleanser over the face and neck.

   ____

3. Describe what is used to remove the cleanser.

   ____

4. Identify the area that requires an especially gentle touch. ____________

5. Specify which types of lotions are applied to oily and dry skins.

   Oily: ____________

   Dry: ____________

6. Explain when to apply a moisturizing lotion. ____________

7. Explain how to test the color of the foundation.

   ____

8. Describe how to apply foundation.

   ____

   ____

9. Explain what is used to remove excess foundation.

   ____

10. Describe how to give the face a matte look.

    ____

11. Explain the purpose of having the client smile. ____________

12. Specify when to apply powdered cheek color. ______________________

13. Describe how to minimize a feature.

____

14. Describe how to emphasize a feature.

____

15. Explain how to select eye color. ______________________

16. Identify an area of the eye that may need shading.

____

17. Identify an area of the eye that may need highlighting. ______________________

____

18. Explain two reasons for using eyeliner.

    1. ______________________ 2. ______________________

19. Explain how to select the color for the eyeliner.

____

20. Describe how to apply eyeliner. ______________________

____

21. Describe the strokes needed to apply eyebrow makeup. ______________________

22. Specify what is used to apply eyebrow makeup. ______________________

23. Specify where to apply mascara. ______________________

____

24. Describe the strokes needed to apply mascara. ______________________

25. Specify what is used to separate the lashes. ______________________

26. Explain how to remove lip color (lipstick) from its container. ______________________

27. Explain how the cosmetologist can steady his/her hand.

____

28. Describe the position of the client's lips when lip color is brushed on.

____

29. Describe the position of the client's lips that makes it possible to smooth lip color into crevices. ______________________

30. List two reasons for blotting lips with tissue.

    1. ______________________ 2. ______________________

31. List two reasons why powdering over the lips is not recommended.

    1. ______________________ 2. ______________________

MAKEUP TECHNIQUES FOR THE BLACK WOMAN

32. List three steps to be completed before applying makeup.

    1. ______________________ 2. ______________________

    3. ______________________

PROCEDURE FOR MAKEUP APPLICATION

33. List the nine steps in the makeup application for the black client.

    1. ______________________
    2. ______________________
    3. ______________________
    ____
    4. ______________________
    5. ______________________
    ____
    6. ______________________
    7. ______________________
    8. ______________________
    9. ______________________

FACIAL FEATURES

34. List three types of subtle facial imperfections.

    1. ______________________
    2. ______________________
    3. ______________________

35. Explain the role of facial makeup when dealing with subtle imperfections.

    ____

ANALYZING THE CLIENT'S FACIAL FEATURES AND FACE SHAPE

36. Identify those features that need emphasizing. ______________________

37. Identify those features that need minimizing.

    ____

38. Name two ways to become more proficient at applying makeup.

    1. ______________________ 2. ______________________

39. Identify the face shape that is considered the ideal.

    ____

40. Specify what needs to be enhanced. ____________________

CORRECTIVE MAKEUP TECHNIQUES

41. Describe how the three lengthwise divisions of the face are measured.

    First third: ____________________

    Second third: ____________________

    Last third: ____________________

42. Describe the proportions of the ideal oval face.

    ____

43. Describe the round-shaped face.

    ____

    ____

44. Explain the aim of corrective makeup for the round-shaped face.

    ____

45. Describe the square-shaped face.

    ____

46. Explain the aim of corrective makeup for the square-shaped face.

    ____

47. Describe the pear-shaped face.

    ____

48. Explain the aim of corrective makeup for the pear-shaped face.

    ____

49. Describe the heart-shaped face. ____________________

50. Explain the aim of corrective makeup for the heart-shaped face.

    ____

51. Describe the diamond-shaped face.

    ____

52. Explain the aim of corrective makeup for the diamond-shaped face.

    ____

53. Describe the oblong-shaped face.

    ____

    ____

54. Explain the aim of corrective makeup for the oblong-shaped face.

    ____

    ____

TIPS FOR APPLYING CHEEK COLOR

55. Explain what is accented by applying cheek color.

____

56. List the four general rules for cheek color application.
    1. ____

       ____
    2. ____
    3. ____
    4. ____

       ____

BASIC RULES FOR APPLYING CORRECTIVE BASE MAKEUP

57. Specify how to accent facial features. ____
58. Specify how to subdue facial features. ____
59. Specify how to balance facial features. ____
60. Explain how to produce a highlight.

____

61. Explain how to form a shadow. ____
62. Describe what the use of shadows does to prominent features.

____

63. Explain why two different shades of foundations must be blended carefully.

____

64. Explain how to achieve color harmony.

____

65. List the three ways to determine a client's most flattering colors.
    1. ____
    2. ____
    3. ____

       ____

66. Specify what is used to conceal age lines and wrinkles due to dry skin.

____

67. Explain the purpose of applying a lighter foundation to a low forehead.

____

68. Explain the purpose of applying a darker foundation to a bulging forehead.

69. Explain the reason for applying a darker foundation on a large nose and a lighter foundation on the cheeks at the sides of the nose.

70. Explain the purpose of applying a light foundation down the center of a short, flat nose.

71. Explain how to minimize a broad nose.

72. Explain how to correct a protruding chin and receding nose.

73. Explain how to highlight a receding (small) chin.

74. Explain how to correct a sagging double chin.

75. Explain the reason for applying a darker foundation over the heavy area of broad jaws.

76. Explain how to highlight a narrow jawline.

77. Explain the purpose of applying a darker shade of foundation over the prominent area of the jawline of a round, square, or triangular face.

78. Explain how to make a short, thick neck appear thinner.

79. Explain the reason for applying a lighter foundation on a long, thin neck than the one used on the face.

80. Explain how to lengthen round eyes.

81. Explain how to correct close-set eyes.

82. Explain how to minimize bulging eyes.

83. Explain how to correct heavy-lidded eyes.

84. Explain how to make small eyes appear larger.

85. Explain how to correct eyes set too far apart.

86. Explain how to correct deep-sunken eyes.

87. Explain how to correct dark circles under eyes.

THE USE OF THE EYEBROW PENCIL

88. Explain what to do when there are spaces in the brow devoid of hair.

89. Explain what to do when the arch is too high.

90. Explain what to do when the arch is too low.

91. Explain why an arch that is too thin and too round is to be avoided when the client has a high forehead.

92. Explain the effect of a low arch on a low forehead.

93. Explain what happens when the eyebrow lines are extended to the inside corners of the eyes.

94. Explain how to make close-set eyes appear farther apart.

____

95. Explain how to make a round face appear narrower. ____________________

96. Explain how to make a long face appear shorter.

____

97. Explain how to offset a narrow forehead.

____

98. Explain how to make a square face appear more oval.

____

99. Identify what the natural arch of the eyebrows follows.

____

100. Specify what needs to be removed to give a clean and attractive appearance.

____

101. Name two alternatives to tweezing for clients who cannot tolerate tweezing.

1. ____________________ 2. ____________________

102. List the implements, supplies, and materials needed for eyebrow arching.

1. ____________________ 2. ____________________
3. ____________________ 4. ____________________
5. ____________________ 6. ____________________
7. ____________________ 8. ____________________
9. ____________________

103. List two ways of seating the client.

1. ____
2. ____

104. Specify what to discuss with the client.

____

105. Describe what is used to cover the client's eyes.

____

106. Explain the purpose of brushing the brows with a small brush.

____

107. Describe what is placed over the brows to soften them.

____

108. Describe how to hold the skin to stretch it when tweezing.

____

109. Identify the number of hairs to grasp with tweezers.

____

110. Describe how to pull the hair.

____

111. Explain how to avoid infection.

____

112. Explain why the hairs between and above the brow line are tweezed first.

____

113. Specify the direction in which to brush the hair when removing hairs from above the brow line. ______________________________

114. Specify the direction in which to brush the hair when removing hairs from under the brow line. ______________________________

115. Explain how to contract the skin after tweezing is completed.

____

116. Specify how often the brows should be treated. ______________________________

LASH AND BROW TINT

117. Explain why aniline derivative tint should never be used to color eyebrows or eyelashes.

____

118. Explain when to use black coloring and when to use brown.

____

____

119. Identify what to follow when applying the coloring agent.

____

120. List the materials and implements needed for a lash and brow tint.

1. ______________________________
2. ______________________________
3. ______________________________
4. ______________________________
5. ______________________________
6. ______________________________
7. ______________________________
8. ______________________________
9. ______________________________

121. Specify what kinds of measures must be followed during the lash and brow tint.

____

122. Explain why the client should not be permitted to lie in a straight position.

____

123. Identify what is used to wash the lashes and brows.

____

124. Name area on which to apply petroleum jelly. ______________________

125. Explain how to adjust the eye shields.

____

126. Specify what is used to apply No. 1 solution to the lashes. ______________________

127. Explain how to remove excess moisture. ______________________

128. Describe where on the lashes the No. 1 solution is applied.

____

129. Specify how often to moisten the lashes. ______________________

130. Explain when to use a fresh applicator stick. ______________________

131. Describe how to apply the No. 1 solution to brows.

____

132. Explain what happens when bottle caps are interchanged.

____

133. Explain how to apply No. 2 solution to lashes and brows.

____

134. Explain what to do if stain gets on the skin. ______________________

135. Describe how to wash lashes and brows after removing eye shields.

____

136. Describe what to do to the brows after placing eye pads over the eyelids.

____

137. Describe what to do to the lashes after placing a small roll of cotton under them.

____

138. Identify the solution used to remove stains. ______________________

139. Specify what is used to soothe the skin. ______________________

140. Identify what is used to wash the eyes. ______________________

141. Name the final step in the procedure for lash and brow tint.

____

ARTIFICIAL EYELASHES

142. List two reasons why a client may want to wear artificial lashes.

    1. ____________________

    2. ____________________

143. Name two basic types of artificial eyelashes.

    1. ____________________

    2. ____________________

APPLYING STRIP EYELASHES

144. Name three materials used to make strip eyelashes.

    1. ____________________ 2. ____________________

    3. ____________________

145. Explain why synthetic fiber eyelashes do not react to changes in weather conditions.

    ____

146. Name the two most popular colors for strip eyelashes.

    1. ____________________ 2. ____________________

147. List the equipment, implements, and materials needed to apply strip eyelashes.

    1. ____________________
    2. ____________________
    3. ____________________
    4. ____________________
    5. ____________________
    6. ____________________
    7. ____________________
    8. ____________________
    9. ____________________
    10. ____________________
    11. ____________________
    12. ____________________
    13. ____________________
    14. ____________________
    15. ____________________
    16. ____________________

148. Name the first step in the procedure for applying strip eyelashes.

____

149. Name the second step in the procedure for applying strip eyelashes.

____

150. Describe how to position the client for application of strip eyelashes.

____

151. Explain what to avoid when lighting the client's face.

____

152. Describe how to ensure that the lash adhesive will adhere properly.

____

153. Identify what must be removed before the procedure is started. ____________

154. Explain the purpose of brushing the client's eyelashes.

____

155. Describe what can be done to straight eyelashes before applying artificial lashes.

____

156. Name two things to be discussed with the client before beginning the application.

1. ____________________ 2. ____________________

157. Describe the position of the cosmetologist when applying artificial eyelashes.

____

158. Specify what to follow during the eyelash application.

____

159. Explain why it may be necessary to trim the outside edge of the upper lash.

____

160. Describe how to make the eyelash strip more flexible so it fits the contour of the eyelid.

____

161. Explain how to feather the lash to create a more natural look.

____

162. Describe the amount of lash adhesive to apply to the base of the lash. ____________

163. Explain where to place the shorter (inside) part of the lash. ____________

164. Describe where to position the rest of the lash.

____

165. Identify what is used to press the lash on. ______________________

166. Explain what to do if eyeliner is used.

____

____

167. Describe how to position the lower lash. ______________________

168. Identify where to place the shorter lash. ______________________

169. Specify where to place the longer lash. ______________________

170. Name two things that will loosen artificial lashes.

1. ______________________ 2. ______________________

REMOVING ARTIFICIAL STRIP EYELASHES

171. Describe a commercial preparation that can be used to remove strip eyelashes.

____

172. Describe a way to soften the lash base.

____

173. Specify for how long to hold the pad or cloth over the eyes to soften the adhesive.

____

174. Identify which corner of the lashline at which to start removing the lashes.

____

175. Explain how to avoid pulling out the client's natural lashes.

____

APPLYING SEMI-PERMANENT INDIVIDUAL EYELASHES (EYE TABBING)

176. Define eye tabbing.

____

177. Explain why synthetic fibers are used in the manufacture of false eyelashes.

____

178. Describe how synthetic lashes are attached.

____

179. Explain why synthetic lashes are referred to as "semi-permanent eyelashes."

____

180. Explain why the false lashes should be filled in by periodic visits to the salon.

____

181. List the items needed to apply individual eyelashes.

1. ______ 2. ______
3. ______ 4. ______
5. ______ 6. ______
7. ______ 8. ______
9. ______ 10. ______
11. ______ 12. ______
13. ______ 14. ______
15. ______ 16. ______
17. ______

182. Explain why it is advisable to give an allergy test before applying the lashes.

______

183. Describe two methods of giving the allergy test.

1. ______ 2. ______

184. Explain how to determine if it is safe to proceed with the application.

______

185. List the four lengths of false eyelashes.

1. ______ 2. ______
3. ______ 4. ______

186. Describe how to create a natural effect with false eyelashes.

______

187. Describe how to create a luxurious effect with false eyelashes.

______

188. Describe how to create a very glamorous or high styling effect with false eyelashes.

______

189. Describe how to create special effects with false eyelashes.

______

190. Specify what the cosmetologist must do before beginning the application of false eyelashes. ______

191. Describe how to position the client for the application of false eyelashes.

______

192. Describe how to light the client's face.

____

193. Explain why adhesive may not adhere properly.

____

194. Name two reasons for brushing the client's lashes.

1. ____
2. ____

195. Describe the effect that false eyelashes can achieve.

____

196. Describe the position of the cosmetologist when applying false eyelashes.

____

197. Explain why only a small quantity of adhesive should be used for each lash.

____

198. Identify what is used to remove the eyelash from the tray. ______________________

199. Explain how to hold the lash. ______________________________

200. Specify when to move the tweezer past the center of the lash.

____

201. Explain how to apply adhesive to the individual lash.

____

202. Explain what to do if too much adhesive is picked up.

____

203. Explain what to do if the lash applied to the center of the eyelid touches the client's glasses. ____

204. Describe the angle at which the lash is held in the tweezers.

____

205. Specify where right-handed cosmetologists should begin applying lashes.

____

206. Specify where left-handed cosmetologists should begin applying lashes.

____

207. Explain how to give a gradual, more natural buildup to the lashes.

____

____

208. Identify which side of the client's natural lash is brushed with adhesive.

____

209. Specify how much of the natural lash is brushed with adhesive.

____

210. Describe how the individual lash is placed onto the natural lash.

____

____

211. Identify what must be kept off the tweezer for efficient performance. _______________

212. Explain the purpose of gently extending the eyelid and holding it taut.

____

____

213. Explain how to prevent the eyelids from sticking together when attaching lashes in the corners of the eyes.

____

____

214. Describe the position of the cosmetologist when applying lower lashes.

____

215. Describe how the client's eyes must be positioned for application of the lower lashes.

____

216. Specify the length of the lashes applied to the lower lid. _______________

217. Describe how to apply adhesive to the lower lash.

____

218. Explain why the client's eyes should be kept open for a few extra seconds.

____

219. Explain the reason for using more adhesive when applying lower lashes.

____

220. Explain why lower lashes will not stay on as long as upper lashes.

____

CORRECTIVE MAKEUP FOR LIPS

221. Describe the usual proportion of the lips.

____

222. List four variations in lips.

1. ____________________ 2. ____________________

3. ____________________ 4. ____________________

FACIAL MAKEUP

223. Create the illusion of better proportions, using a red pencil, on the illustrated lips.

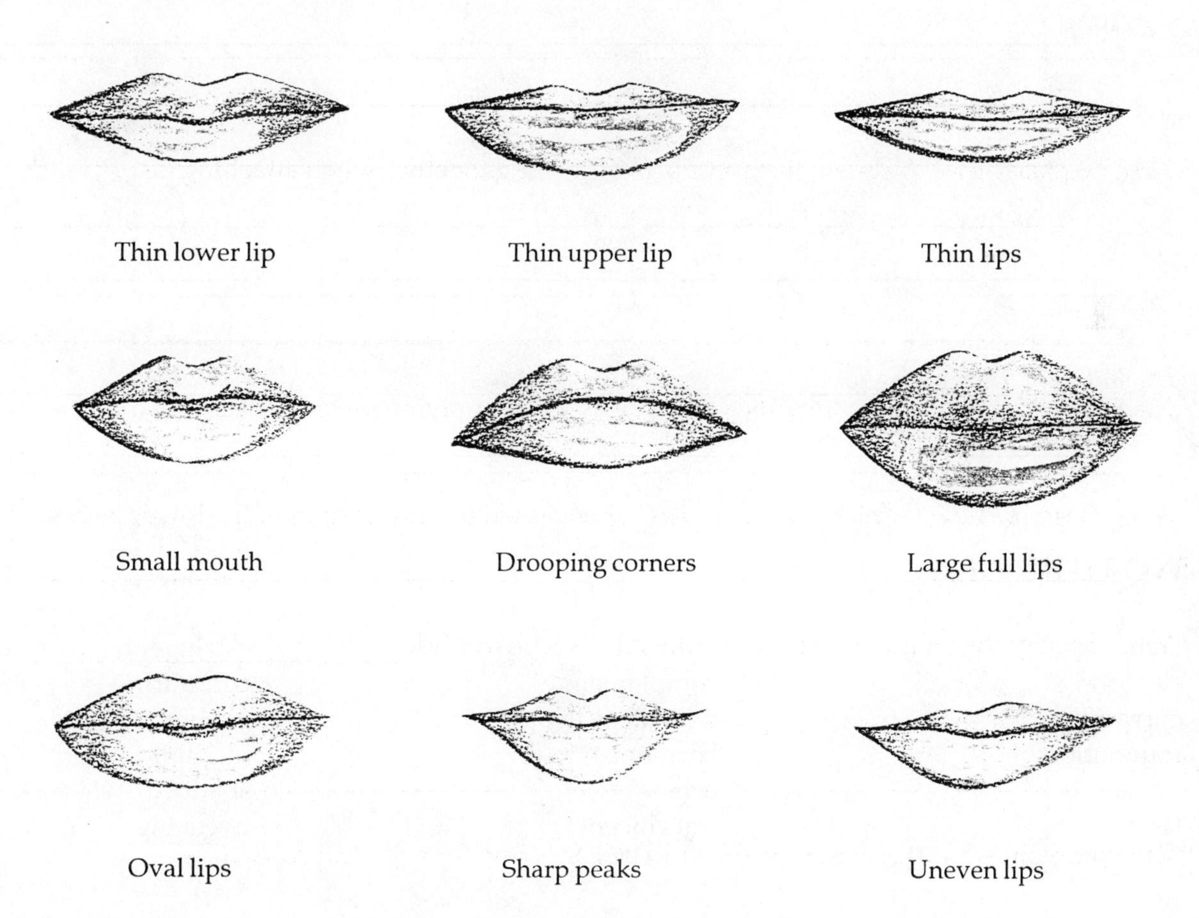

SAFETY PRECAUTIONS

224. List the safety precautions associated with the application of makeup and eyelashes.

1. ______________________________

   ____

2. ______________________________

   ____

3. ______________________________
4. ______________________________
5. ______________________________
6. ______________________________
7. ______________________________
8. ______________________________
9. ______________________________

   ____

10. ______________________________
11. ______________________________
12. ______________________________
13. ______________________________

    ____

14. ______________________________

    ____

15. ______________________________

WORD REVIEW

arch
cleanser
corrective
foundation
moisturizer
toner

artificial
complement
depilatory eye
harmonize
orbit
translucent

astringent
contouring
tabbing
highlight
shadow
tweezing

MATCHING TEST

Insert the correct term or phrase in front of each definition.

allergy test	astringent	color harmony
eye tabbing	foundation cream	highlight
lash and brow tint	moisturizer	shadow
skin freshener		

1. ________________________the technique of attaching individual, synthetic eyelashes to a client's own eyelashes
2. ________________________an effect produced when the foundation used is darker than the original color
3. ________________________a lotion applied to dry skin after it is cleansed
4. ________________________a procedure given to determine if the client is sensitive to the eyelash adhesive
5. ________________________an effect achieved when the makeup tones flatter the client's eyes, hair, and skin
6. ________________________a lotion applied to dry, delicate skin before foundation is applied
7. ________________________a product used to conceal age lines and wrinkles due to dry skin
8. ________________________an effect produced when a shade lighter than the original foundation is used
9. ________________________a lotion applied to oily skin after it is cleansed
10. ________________________a harmless coloring agent for eyelashes and eyebrows

RAPID REVIEW TEST

Place the correct word in the space provided in each sentence below.

adhesive	blindness	center
cheek	clothes	complement
container	corner	curl
darker	demarcation	eyebrows
eyeliner	fingernails	hands
highlighting	lighter	lip
orbit	oxidation	prominent
shadowing	spatula	top

1. Cosmetics are removed from their containers with a ______________ or cosmetic applicator.
2. To assure a more lasting application of lower lashes, more ______________ is needed than when applying upper lashes.
3. When two shades of foundation are used, care must be taken to blend them so that there will be no line of ______________
4. Round eyes can be lengthened by extending the shadow beyond the outer ______________ of the eye.
5. The natural arch of the eyebrow follows the curved line of the ______________.
6. Cosmetologists can avoid scratching the client's skin by keeping ______________ smooth.
7. A sagging double chin can be minimized by applying a ______________ foundation to the sagging portion.
8. ______________ is used to make the eyes appear larger and the lashes thicker.
9. If bottle caps of solutions for lash and brow tint are interchanged, ______________ starts and the liquids lose their value.
10. Close-set eyes can be made to appear farther apart by widening the distance between the ______________
11. Facial features are emphasized by ______________ and minimized by ______________
12. Synthetic-fiber eyelashes are made with a permanent ______________.
13. Eye color should match or ______________ the eyes.
14. Using an aniline derivative tint on eyelashes and/or eyebrows can cause ______________

15. Bulging eyes can be minimized by blending a dark shadow carefully over the ________________ part of the upper lid.

16. Lip color must not be applied to the client's lips directly from the ________________

17. Individual eyelashes are placed on __________ of the natural lashes.

18. ____________ color should not be applied in a bright, round circle.

19. A ____________ foundation cream helps to broaden the forehead between the brows and hairline.

20. Cosmetologists must wash and sanitize their ____________ before and after every makeup application.

MULTIPLE CHOICE TEST

Read each statement carefully, then write the letter representing the word or phrase that correctly completes the statement on the blank line to the right.

1. The first step in a professional makeup application is to apply
   a) foundation b) corrective makeup
   c) cleansing cream or lotion d) astringent lotion ________

2. A product that should never be used on eyelashes and/or eyebrows is
   a) mascara b) eyelash adhesive
   c) aniline derivative tint d) lash and brow tint ________

3. The purpose of corrective makeup for the diamond-shaped face is to reduce the width across the
   a) forehead b) temples
   c) cheekbone line d) jawline ________

4. When tweezing hair from under the eyebrow line, the hairs are brushed
   a) upward b) downward
   c) toward the right side d) toward the left side ________

5. Immediately following the makeup application, used pencils and applicators must be
   a) rinsed b) washed
   c) sanitized d) discarded ________

6. Extending the eye shadow slightly above, beyond, and below the eyes makes eyes appear
   a) smaller b) larger
   c) closer together d) more bulging ________

7. Natural oils from the eyelids tend to dissolve the
   a) artificial eyelashes b) mascara
   c) eyelash adhesive d) eyebrow pencil ________

8. The color of the foundation is tested by blending it on the client's
   a) forehead b) nose
   c) cheeks d) jawline ________

9. When tweezing eyebrows, infection can be avoided by applying
   a) an antiseptic b) a moisturizer
   c) a styptic d) an eyebrow tint ________

10. The face shape considered to be ideal is
    a) round-shaped b) diamond-shaped
    c) pear-shaped d) oval-shaped ________

11. Individual eyelashes usually stay on for about
    a) 6 to 8 hours b) 6 to 8 days
    c) 6 to 8 weeks d) 6 to 8 months ________

12. For the application of lash and brow tint, the client should be
    a) standing up b) sitting up straight
    c) partially reclining d) lying in a straight position ________

13. A moistened cosmetic sponge may be pressed over the finished makeup to give the face a
    a) shiny look b) matte look
    c) glossy look d) lustrous look ________

14. The application of artificial eyelashes may proceed if there is no reaction from the allergy test within
    a) 6 hours b) 12 hours
    c) 18 hours d) 24 hours ________

15. Hairs between the brows and above the brow line are tweezed first, because the area under the brow line is much more
    a) hairy b) relaxed
    c) oily d) sensitive ________

16. After the application of lip color, the lips are blotted with
    a) tissue b) cotton
    c) a towel d) a drape cloth ________

17. Cosmetologists must protect the client's eyes with shields when applying
    a) eye makeup b) lash and brow tint
    c) artificial strip eyelashes d) individual eyelashes ________

18. Eyebrow pencil is applied in strokes that are
    a) bold b) heavy
    c) light and feathery d) dark and feathery ________

19. Individual eyelashes are applied with
    a) the fingers b) a spatula
    c) a cotton-tipped swab d) a tweezer ________

20. Eyebrows that are too thin and too round give the client a

a) sad look
b) frowning look
c) surprised look
d) tired look ________

21. The oblong-shaped face is

a) short and narrow
b) long and narrow
c) short and wide
d) long and wide ________

22. When applying artificial eyelashes, the eyelids are prevented from sticking together by separating the upper and lower lashes for

a) a second
b) several seconds
c) several minutes
d) several hours ________

23. Powdering over the lips after applying lip color removes the

a) dry look
b) chapped look
c) caked look
d) moist look ________

24. The client's hair and clothing are protected during the makeup procedure by

a) cleansing
b) draping
c) moisturizing
d) concealing ________

25. During tweezing, eyebrow hairs are grasped individually with tweezers and pulled with a

a) quick motion
b) hesitant motion
c) jerky motion
d) slow motion ________

26. When lash and brow tint gets on the skin, it must be removed immediately with

a) hair lightener
b) stain remover
c) skin freshener
d) moisturizing lotion ________

27. Towels, linens, and makeup cape should be placed in the proper container until they can be washed and

a) rinsed
b) dried
c) ironed
d) sanitized ________

28. Whether working with artificial eyelashes or lash and brow tint, cosmetologists must follow the directions of

a) the client
b) the manufacturer
c) a more experienced cosmetologist
d) the salon manager ________

29. Eyebrows should be tweezed about once every

a) day
b) week
c) month
d) year ________

30. Lips are usually proportioned so that the curves or peaks of the upper lip fall directly in line with the

a) chin
b) upper teeth
c) nostrils
d) bridge of the nose ________

Also see *Milady's Standard Theory Workbook.*

THE SKIN AND ITS DISORDERS

See *Milady's Standard Theory Workbook.*

Date ____________________

Rating ____________________

Text Pages 451–460

REMOVING UNWANTED HAIR

TREATMENT

1. List the equipment, implements, and materials needed for thermolysis.

   1. ____________________
   2. ____________________
   3. ____________________
   4. ____________________
   5. ____________________
   6. ____________________
   7. ____________________
   8. ____________________
   9. ____________________
   10. ____________________
   11. ____________________
   12. ____________________
   13. ____________________
   14. ____________________
   15. ____________________
   16. ____________________
   17. ____________________

2. Give another name for thermolysis epilators. ____________________

3. List the nine control systems that all thermolysis epilators have in common.

   1. ____________________ 2. ____________________
   3. ____________________ 4. ____________________
   5. ____________________ 6. ____________________
   7. ____________________ 8. ____________________
   9. ____________________

4. Name two innovations that some manufacturers have incorporated into the control systems of their machines.

   1. ____
   2. ____

5. Explain why it is necessary to follow the instruction manual provided by each machine's manufacturer. ____

6. Specify what should be of utmost importance to the electrologist.

   ____

7. Explain why the comfort of the electrologist is also important.

   ____

PROCEDURE FOR THERMOLYSIS

8. List the 12 steps required in the procedure for thermolysis.

   1. ____
   2. ____
   3. ____
   4. ____
   5. ____
   6. ____
   7. ____
   8. ____
   9. ____
   10. ____
   11. ____
   12. ____

9. Identify what is the most critical technique in electrology treatment.

   ____

10. Explain what determines the size of the needle.

    ____

11. Describe how to hold the needle holder.

    ____

12. Explain what the angle of insertion depends on.

    ____

13. Describe how to insert the needle for a hair that grows straight up.

    ____

14. Explain what to do when there is an excessive amount of hair and the hairs are long.

____

15. Identify what is used to gently remove the hair. _______________

16. Explain what to do once the current is turned off.

____

17. Explain how to hold the needle holder and tweezer.

____

18. Explain how to hold the area being treated.

____

____

19. Identify the four skills that have to be developed to increase treatment efficiency.

1. _______________ 2. _______________

3. _______________ 4. _______________

20. Specify what to do when the treatment is completed. _______________

21. Describe how to soothe the skin and aid in the healing process.

____

____

ELECTRONIC TWEEZER METHOD OF HAIR REMOVAL

22. Name another method of removing superfluous hair used in salons.

____

23. Describe where the radio frequency energy is transmitted.

____

24. Explain what (supposedly) happens to the papilla.

____

25. Specify how many strands of hair are grasped by the tweezer. _______________

26. Explain why the energy is first applied at a low level. _______________

27. Explain why the energy is then applied at a higher level. _______________

28. Explain what most manufacturers suggest to increase efficiency.

____

29. List two disadvantages of the electronic tweezer.

1. _______________

2. _______________

SHAVING

30. Explain when shaving is recommended. ______________________

31. Name two areas that are commonly shaved.

    1. ______________________ 2. ______________________

32. Specify what is applied before shaving. ______________________

33. Identify an implement also used for shaving. ______________________

34. Explain how to reduce skin irritation. ______________________

35. Specify for what purpose an electric clipper is most often used.

    ____

36. Explain why shaving only seems to make the hair grow thicker or stronger.

    ____

TWEEZING

37. Name two common uses for tweezing.

    1. ______________________
    2. ______________________

DEPILATORIES

38. Name two types of depilatories.

    1. ______________________ 2. ______________________

HOT WAX

39. List six areas of the body on which hot wax may be applied.

    1. ______________________ 2. ______________________
    3. ______________________ 4. ______________________
    5. ______________________ 6. ______________________

40. Explain why lanugo hair should not be removed. ______________________

41. Explain why waxing and tweezing may cause the hair to grow stronger.

    ____

42. Specify two methods for melting the wax.

    1. ______________________ 2. ______________________

43. Describe what is used to wash the skin. ______________________

44. Identify what is spread over the skin surface after it is rinsed and dried.

    ____

45. Explain how to test the temperature and consistency of heated wax.

____

46. Specify what is used to spread the warm wax over the skin. ______________

47. Describe the direction in which to spread the warm wax over the skin.

____

48. Explain how to apply a sanitized cloth strip.

____

49. Describe what the wax must do before it is pulled off. ______________

50. Identify the direction in which to pull off the adhering wax.

____

51. List the three steps required to complete the procedure.

1. ______________________________
2. ______________________________
3. ______________________________

52. Explain how to prevent burns to the client's skin.

____

53. Identify an area the wax should not contact. ______________

54. Name six areas to which a wax depilatory should not be applied.

1. ______________ 2. ______________
3. ______________ 4. ______________
5. ______________ 6. ______________

COLD WAX

55. Identify what is available for clients who cannot tolerate heated wax. ______________

56. Compare cold wax to hot wax.

____

____

57. Identify the temperature of cold wax when it is applied. ______________

58. Explain how to spread a thin coat of wax.

____

59. Describe how to make wax adhere to a strip of cellophane or cotton cloth.

____

60. Describe how to hold the skin to pull off the wax strip. ______________

61. Specify what kind of movement is required to pull off the wax strip.

____

62. Name the direction required in which to pull off the wax strip.

____

CHEMICAL DEPILATORIES

63. List the three forms of chemical depilatories.

1. ____________________ 2. ____________________
3. ____________________

64. Name the area on which chemical depilatories are generally used. ____________

65. Explain the purpose of the skin test.

____

____

66. Specify the area where the skin test is given. ____________________

67. Explain what to follow when giving a skin test. ____________________

68. Specify how long to leave the depilatory on the skin. ____________________

69. Explain how to determine if the depilatory can be used safely over a large area of skin.

____

70. Describe how to apply the cream type of depilatory. ____________________

71. Describe how to mix the powder type. ____________________

72. Explain how to apply the depilatory after cleansing and drying the skin.

____

73. Describe how to protect the surrounding skin. ____________________

74. Specify the length of time to keep depilatory on the hair. ____________________

75. Explain how hair thickness determines processing time.

____

76. Specify the water temperature used to remove the depilatory. ____________________

77. Identify what to apply to the skin after patting it dry. ____________________

WORD REVIEW

chemical
cold wax
depilatory
electrologist
electrology
electronic tweezer
epilator
follicle
hot wax
lanugo
papilla
physical shaving
shortwave
thermolysis
tweezing

MATCHING TEST

Insert the correct term or phrase in front of each definition.

chemical depilatory
shaving
tweezing
electrology
shortwave
wax depilatory
electronic tweezer
skin test

1. ______________________ a temporary method of hair removal requiring a skin test before being used over a large area
2. ______________________ a temporary method of hair removal commonly used for shaping the eyebrows
3. ______________________ the permanent method of hair removal
4. ______________________ a procedure that determines whether an individual is sensitive to a particular product
5. ______________________ a temporary method of hair removal recommended for the legs and under arms
6. ______________________ another name for thermolysis
7. ______________________ a temporary method of removing hair by applying it either heated or cold
8. ______________________ a temporary method of hair removal in which radio frequency energy is transmitted down the hair shaft into the follicle area

RAPID REVIEW TEST

Place the correct word in the space provided in each sentence below.

angle
cold
hair
manufacturer
papilla
redness
skin
tweezer
chemical
epilators
holder
needle
paste
resistance
slow
clipper
follicle
hot
opposite
physical
same
strip

1. Because of variations in thermolysis epilators, electrologists are advised to follow the instruction manual provided by each ___________________.

2. Wax depilatories are applied over the skin surface in the ___________ direction as the hair growth.

3. The powder type of chemical depilatory is mixed with water to form a smooth ___________.

4. In electrology, the diameter of the hair being treated determines the size of the ___________.

5. Hot and cold waxes are ______________ types of depilatories.

6. The claim associated with the electronic tweezer method is that it dehydrates and eventually destroys the ____________.

7. A pre-shave lotion helps to reduce skin irritation from an electric ____________.

8. All thermolysis ______________ have the same control systems.

9. The thicker the hair, the longer the ______________ depilatory is kept on.

10. In electrology, the angle at which the needle is inserted depends on the ___________ at which the hair is growing.

11. Talcum powder is spread over the skin surface before applying __________ wax.

12. Efficiency is increased with the electronic _____________ by first steaming the area to be treated.

13. Shaving does not cause the __________ to grow thicker or stronger.

14. A chemical depilatory may be used safely if the skin test shows no sign of _____________ or swelling.

15. After applying hot or cold wax, a cloth ___________ is pressed over the wax.

16. When a hair grows straight up, the thermolysis needle is inserted straight down into the ___________.

17. The process of clearing any area of hair with the electronic tweezer method is ___________.

18. In electrology, the needle is inserted until slight _______________ is felt.

19. Wax depilatories are pulled off the skin surface in the ______________ direction as the hair growth.

20. In electrology, the needle ____________ is held in the same way as a pen or pencil.

MULTIPLE CHOICE TEST

Read each statement carefully, then write the letter representing the word or phrase that correctly completes the statement on the blank line to the right.

1. The most critical technique in electrology treatment is
   a) holding the needle
   b) setting the timer
   c) inserting the needle into the follicle
   d) determining the needle size ________

2. Waxing and tweezing may stimulate circulation and increase
   a) muscle tone
   b) blood supply
   c) the number of sweat glands
   d) the number of oil glands ________

3. When using a chemical depilatory, the surrounding skin is protected with
   a) astringent lotion
   b) rubbing alcohol
   c) petroleum jelly or cream
   d) pre-shaving lotion ________

4. An electric clipper is most often used to remove unwanted hair from the
   a) nape
   b) forehead
   c) cheeks
   d) eyebrows ________

5. Removing lanugo hair may cause the skin to lose its
   a) softness
   b) elasticity
   c) color
   d) fat cells ________

6. After electrology treatments, the healing process of the skin is helped by the application of
   a) rubbing alcohol
   b) hydrogen peroxide
   c) an after-treatment lotion
   d) a pre-shaving lotion ________

7. Cold wax removes hair in the same manner as
   a) shaving
   b) tweezing
   c) a chemical depilatory
   d) hot wax ________

8. To help the electrologist see the angle of hair growth more clearly, excessively long or thick hair can be
   a) shampooed
   b) blow-dried
   c) cut
   d) shaved ________

9. Undesirable hairs from around the mouth and chin are best removed by
   a) tweezing
   b) clipping
   c) singeing
   d) shaving ________

10. Chemical depilatories are kept on the hair for
    a) 1 to 3 minutes
    b) 5 to 10 minutes
    c) 15 to 20 minutes
    d) 25 to 30 minutes ________

11. Before an electrology treatment, the needle and tweezers are
    a) blown off
    b) rinsed off
    c) washed off
    d) sterilized ________

12. Wax depilatories should not be used on
    a) the chin
    b) the upper lip
    c) the legs
    d) irritated or inflamed skin ________

13. Shaving blunts the hair ends, causing them to

a) feel stiff
b) feel soft
c) lose color
d) lose elasticity ________

14. Before a chemical depilatory, a skin test is done

a) behind the ear
b) on the upper lip
c) on a hairless part of the cheek
d) on a hairless part of the arm ________

15. Wax depilatories are removed from the skin by

a) chipping off the wax
b) melting the wax
c) pulling off the wax slowly
d) pulling off the wax quickly ________

Also see *Milady's Standard Theory Workbook*

CELLS, ANATOMY, AND PHYSIOLOGY

See *Milady's Standard Theory Workbook.*

ELECTRICITY AND LIGHT THERAPY

See *Milady's Standard Theory Workbook.*

CHEMISTRY

See *Milady's Standard Theory Workbook.*

THE SALON BUSINESS

See *Milady's Standard Theory Workbook.*